DISCARDED

AN ADVENTURE IN MULTIDIMENSIONAL SPACE

The Art and Geometry of Polygons, Polyhedra, and Polytopes

KOJI MIYAZAKI

College of Liberal Arts
Kobe University, Japan
(Asakura Shoten Publishing Company, 1983)

Translated by the Author
Edited and Revised by Henry Crapo

A Wiley-Interscience Publication
JOHN WILEY & SONS

New York · Chichester · Brisbane · Toronto · Singapore

Authorized English translation and adaption from the original Japanese language edition titled, かたちと空間 多次元世界の軌跡 (Forms of Space), by Koji Miyazaki.

Copyright © 1983 by Asakura Publishing Company, Ltd.

English language edition:
Copyright © 1986 by John Wiley & Sons, Inc.

Published by John Wiley & Sons, Inc.

All rights reserved. Published simultaneously in Canada.

Reproduction or translation of any part of this work beyond that permitted by Section 107 or 108 of the 1976 United States Copyright Act without the permission of the copyright owner is unlawful. Requests for permission or further information should be addressed to the Permissions Department, John Wiley & Sons, Inc.

Library of Congress Cataloging in Publication Data:

Miyazaki, Kōji, 1940-
   An adventure in multidimensional space.

   "A Wiley-Interscience publication."
   Includes index.
   1. Visual perception. 2. Space (Art) 3. Form (Aesthetics) 4. Polygons. 5. Polyhedra. 6. Polytopes.
I. Title.
N7430.5.M59 1986   701   85-22595
ISBN 0-471-81648-5

Printed in Japan.

Foreword

Koji Miyazaki's book contains magnificent illustrations of energetic formulations and transformations operative in nature's comprehensive coordinate system. The quality of Miyazaki's drawing and coloring is one more manifestation of the Japanese artist's exquisite conceptualization and rendering. The geometrical presentations themselves are those of an experimental scientist-scholar, exciting and comprehensive.

BUCKMINSTER FULLER

Foreword

Koji Miyazaki has a broad knowledge of the history and geometry of symmetrical figures. He skillfully constructs models of them, and makes artistic use of color. I am particularly impressed by his development of the "golden isozonohedra" (Kepler's triacontahedron, Federov's rhombic icosahedron, Bilinski's "second rhombic dodecahedron", and two kinds of rhombohedron, all having for faces the "golden rhombus" whose diagonals are in the ratio $\pi : 1$, where $\pi = 2\cos 36°$). He puts these solids together, face to face, in such a way as to form a honeycomb which can be continued indefinitely to fill the whole Euclidean space. One especially appealing honeycomb of this kind has pentagonal symmetry, in apparent violation of the famous crystallographic restriction.

H. S. W. Coxeter

Preface

In this book we begin a graphic adventure into two-, three-, and four-dimensional fantastic worlds by looking at polygons (or 2-polytopes), polyhedra (or 3-polytopes), and polytopes (or 4-polytopes). Our topics extend over various ages and into many countries with special emphasis on Japan's past.

The stars of the show are Plato and polygons, Kepler and polyhedra, Fuller and polytopes. Plato in ancient Greece of the fourth century B.C. explained in his *Timaeus* that the cosmos is filled with polygons. These ideas met with objections from Aristotle and his camp, but nevertheless had an influence on Euclid. Kepler in seventeenth-century Germany, astonished by Copernicus, constructed a space filled with polyhedra in his *Harmonices Mundi*. Disregarded by Galileo, he nevertheless gave a hint to Newton. Fuller in the United States of today, impressed by Einstein, writes in his *Synergetics* that he imagines the universe filled with polytopes. This puzzles some people, and amuses others, including Coxeter.

Standing at the center of their respective civilizations, these three men used similar methods in order to investigate the macro, medio, and micro universes by themselves, without any telescope or microscope, and without requiring guidance from others.

And, at last, they each found their way to the entrance of four-dimensional space, though they did not describe the scene concretely.

What spectacle does four-dimensional space present? The clues to reveal it are to be found in Plato's polygons, Kepler's polyhedra, and Fuller's polytopes.

For example, in 3-space the three-leaf clover has three thin leaves and opens into a regular triangle. In 4-space such a clover may have four thick leaves and will open into a regular tetrahedron.

Similarly, in 3-space a four-leaf clover has four thin leaves and opens into a square. In 4-space such a clover may have six or eight thick leaves and will open into a cube or regular octahedron, which are mutually dual.

In 3-space, a cherry blossom has five thin petals and blooms in a regular pentagon. So in 4-space, such a blossom may have twelve or twenty thick petals and will open in a regular dodecahedron or icosahedron, both of which are also mutually dual.

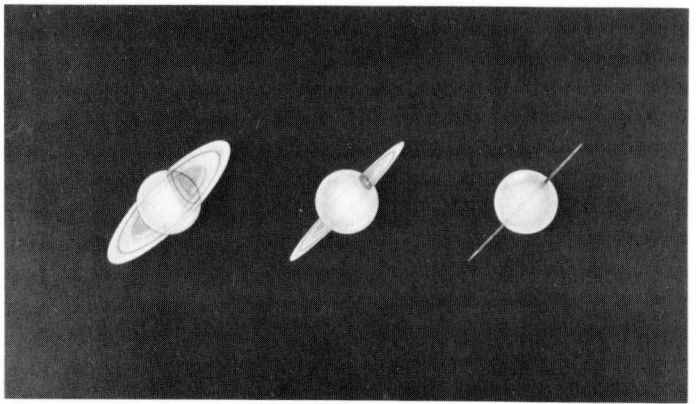

A hyperrainbow, whose shape is not semicircular but semispherical, covers such flowers.

In the high four-dimensional sky above the hyperrainbow, hyper-Saturn sparkles, changing its shape as shown in the pictures at the top of this page. The shape of its ring changes from a line into an ellipsoid, not an ellipse.

In a deep four-dimensional mine hyperdiamonds crystallize into the shape of a hypercube or 16-cell rather than into that of a cube or regular octahedron.

To sum up, polygons, polyhedra, and polytopes are the effective tools or hieroglyphs with which we may investigate and describe the macro, medio, and micro worlds or the multi-dimensional world without any telescope or microscope and without requiring guidance from others.

In acknowledgment of their great contribution to the translation into English from the original Japanese edition of this book, I express my gratitude to Professor Shigeru Maeno, physicist and professor emeritus of English at Kobe University, who is known as the author of *A Melville Lexicon* (in English), and to Professor Henry Crapo, mathematician at the Institut national de recherche en informatique et en automatique (INRIA), Rocquencourt, France, and editor-in-chief of the journal *Structural Topology,* from which I frequently quoted in this book.

Koji Miyazaki

Contents

1	The Circle, by Pythagoras	1
2	The Cosmos, by Plato	8
3	Nature, by Aristotle	16
4	The Family, by Archimedes	21
5	The Dream, by Kepler	28
6	The Mandala, by Kukai	35
7	Synergy, by Fuller	40
8	The Palace, by Kelvin	51
9	The Equation, by Euler	54
10	The Labyrinth, by Möbius	61
11	Magic, by Dürer	69
12	The Planet, by Penrose	76
13	Coordinates, by Descartes	84
14	The Crystal, by Schläfli	87
15	The Creation, by Coxeter	96
16	The Hypersphere, by Einstein	100
	Bibliography	106
	Afterwords	108
	Index	110

The Circle

by Pythagoras

Everything originates in a "big bang"—from a point. Pythagoras gave to the point a circular form which takes pride in its own completeness, like the sun, and he postulated a cosmos composed of a regular arrangement of such equiradial circles. The circle, "En" in Japanese, is closely related to the point, "Ten" in Japanese, and to heaven, also "Ten" in Japanese. According to Pythagoras, a line is determined by two points, or two circles, a plane by three, and a solid or 3-space by four, as in figure [1]. That is all. The total is the perfect number "ten."

Pythagoras christened one form of these ten circles, arranged in a regular triangle on a plane, shown in figure [2], as the tetractys, and suggested this as the form of the macrocosm. For such reasons, medieval theologians compared nature, the macrocosm, to the number ten, and man, the microcosm, to five.

The astronomer and astrologer Kepler said in his *Harmonices Mundi* that, among regular triangular arrangements of circles, the tetractys is the first figure that has a heart at the center and has life as well. If so, why is human life, as the microcosm, symbolized by the number five? Nature is not so simple for us to be able to put it in order by such elementary concepts as the numbers five and ten.

Pythagoras was aware of this inconsistency. He devised quadrangular, pentagonal, and other polygonal numbers in addition to the triangular number, the tetractys, as in figure [3]. He tried to incorporate various natural phenomena in a unified field by using circles. This philosophy, originating in circles and ending in circles, went further than that dominated by straight lines extending to an ominous infinity, and linked the geocentric systems of Plato, Aristotle, Ptolemy, Dante, and others.

Dante, who worried about spending the "unlucky" year 1300 on Earth, wrote in his *Divina Commedia* about the three equiradial circles, the Trinity, seen in the calmness beyond a circular Saturn moving on a circular orbit.

About two centuries later, Copernicus presented the heliocentric system, yet the orbits of the planets were still ruled by Pythagorean circles. After all, the circular restriction by Pythagoras had remained until Kepler found the elliptical orbits of the planets. But, in his innermost heart, Kepler believed

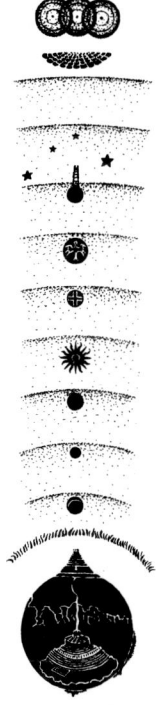

Dante's cosmos in Divina Commedia. *From bottom to top, Earth, the moon, Mercury, Venus, the sun, Mars, Jupiter, Saturn, and fixed stars. The uppermost three equiradius circles are the Trinity.*

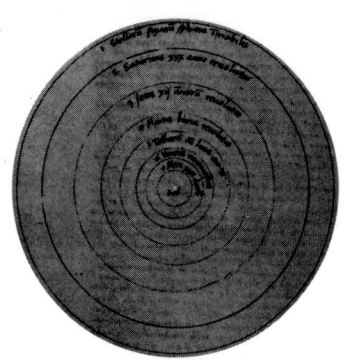

Copernican heliocentric system. The center is the sun, and in the outermost circle, the orbits of Mercury, Venus, Earth, Mars, Jupiter, Saturn, and fixed stars are shown. From Copernicus, De revolutionibus orbium coelestim.

in the Pythogorean circular world, and it is sometimes said that he discovered the elliptical orbits by mistake.

Galileo disregarded his contemporary Kepler, ignoring the elliptical orbits and supporting the heliocentric system with its Pythagorean circles *(Dialogue)*.

Though the Pythagorean thoughts were powerful, there remained some obscure riddles surrounded by the fog of the past. For example, what role do the gaps between the circular disks of the tetractys play in the wholly filled cosmos?

Geometers of today call a group of circles a *packing* of a plane by circles, or more simply a circle arrangement, when each circle is tangent to at least three other circles but leaves gaps, and a *covering* of a plane by circles when the circles overlap each other to cover the whole plane without any gap. The problem is to minimize the area of the gaps in a packing, and that of the overlapping parts in a covering. In the case of equiradial circles, the tetractys, as a packing, and the circle stacking in figure 4, as a covering, are the best answers.

Figure 5, with 11 in all, shows the most typical examples of packings. Some triangular, quadrangular, and pentagonal numbers are also included since these show fundamental models of "molecular structures" in the geometrical and integerlike cosmos of Pythagoras. When the centers of these circles are connected by straight lines passing through the points of tangency, regular polygons are born. Thus the 11 patterns using regular polygons, shown in figure 6, are derived; they are called Pythagorean tessellations. Among them, the three in the top row, which have only one kind of regular polygon fitted together similarly around each vertex, are called regular tessellations, and each of the other eight, which has two or more kinds of regular polygons fitted together similarly around each vertex, is called a semiregular tessellation.

Pythagorean tessellations range over a wide field of design in art and nature. Arabesques which decorate Islamic mosques like honeycombs and also real honeycombs, admired in the *Arabian Nights,* where it is said that Euclid must have studied them, are the typical examples.

In the seventh century, Islamic culture suddenly flourished on a barren tract of land. By contrast, in medieval Europe, people seemed to have fallen asleep in barren darkness, dreaming of the celestial regions where Pythagoras lived as a god. The night sky, after the setting of the hot sun, presented to Arabs a fine spectacle of the paradise twinkling with stars—"Stars lead all," according to the Koran.

The Arabians could not sleep as well as the Europeans. They had to hold the central position of the western civilization for the Europeans, introducing the sacred geometry of ancient Greece and vulgar algebra of ancient India. It is natural that the geometrical and integerlike cosmos of Pythagoras became a religious object for such Islamites. Arabesques, making use

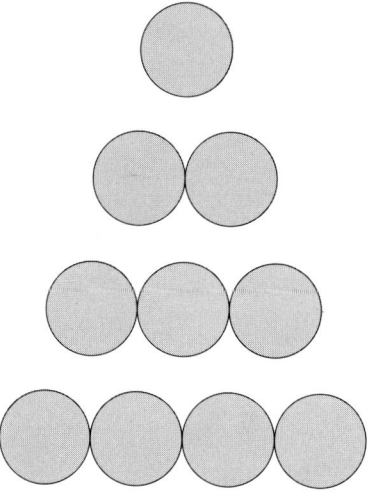

1 Basic construction of the cosmos by Pythagoras. From top to bottom, a point, line, plane, and solid (3-space) represented respectively by equiradial circles.

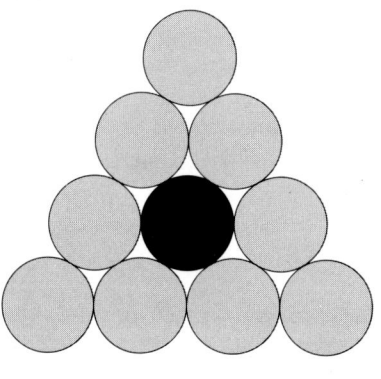

2 Tetractys having a heart.

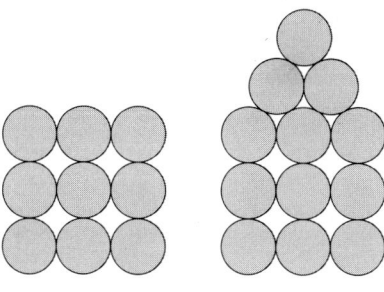

3 Square- (left) and pentagonal number.

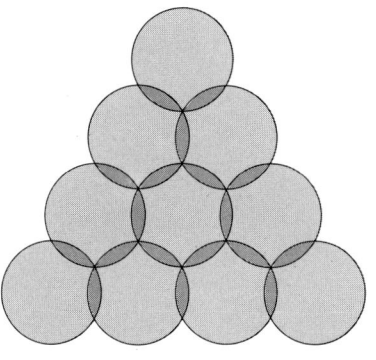

4 Covering of a plane by circles having the least overlapping part.

Stalactite vault like a honeycomb and arabesques. Alhambra.

of Pythagorean tessellations, were so invaluable to them that nongeometrical animal patterns were prohibited.

Critchlow, a modern investigator of mosques, shows arabesques which are derived from each of the Pythagorean tessellations and says that the tetractys is frequently hidden in them. This recalls the Pythagorean cosmos, as shown by pale blue circular discs in the top right of figure ①.

Arabesques brilliantly cover, rather than pack, the inner parts of spherical domes of mosques and represent the heaven filled with stars. Coxeter says that the first person to examine all spherical Pythagorean tessellations was an Arab of the tenth century.

Pythagorean tessellations related to a sphere also appear on the surface of a Japanese artifact, the temari, whose name means a circular or spherical thing, and which has been handed down from about the time when Islam spread throughout Arabia. There is a temari-shaped incense burner in the Shosoin-treasure house in Nara, Japan, which may have come from Arabia in the eighth century. It has an openwork arabesque pattern of twelve spherical pentagons. At that time, a temari was sometimes worshiped as an image of a god.

Figure ② shows temaris devised by a modern craftswoman. In most cases they are decorated with one of three regular spherical polygonal patterns shown in figure ③. The threads follow great circles, whose radii are equal to those of the spheres, so they cannot get loose.

Those three patterns are obtained by equal division of great circles. Modern temari-craftswomen call the central one, having 24 congruent spherical triangles, the hexasect (T_6), the left one, having 48 congruent spherical triangles, the octasect (T_8), and the right one, having 120 triangles, the decasect (T_{10}), according to the number of points of division. To derive T_{10}, one needs a little secret adjustment, using a formula found by a modern geometer who likes temaris.

The patterns of temaris produce polygonal tessellations on a sphere. They hint at a world of polyhedra, rather than polygons.

If the spherical polygons seen in spherical tessellations are replaced by planar polygons, certain well-known convex polyhedra, all of whose dihedral angles (angles subtended by the neighboring faces) are less than 180 degrees, are obtained. Among such convex polyhedra, the most vulgar are the prisms and pyramids, and the most sacred are the regular polyhedra.

Regular polyhedra are convex polyhedra, each of which has only one kind of regular polygon fitted together similarly around each vertex. There are five and only five regular polyhedra, as shown in figure ④. The central one is the regular tetrahedron, having four regular triangles, three of them fitted together at each vertex. Left of center is the regular hexahedron or cube having six squares, three of which are fitted

5 ▶

Three regular (uppermost row) and eight semiregular packings of a plane by circles.

6 ▶

Three regular (uppermost row) and eight semiregular tessellations.

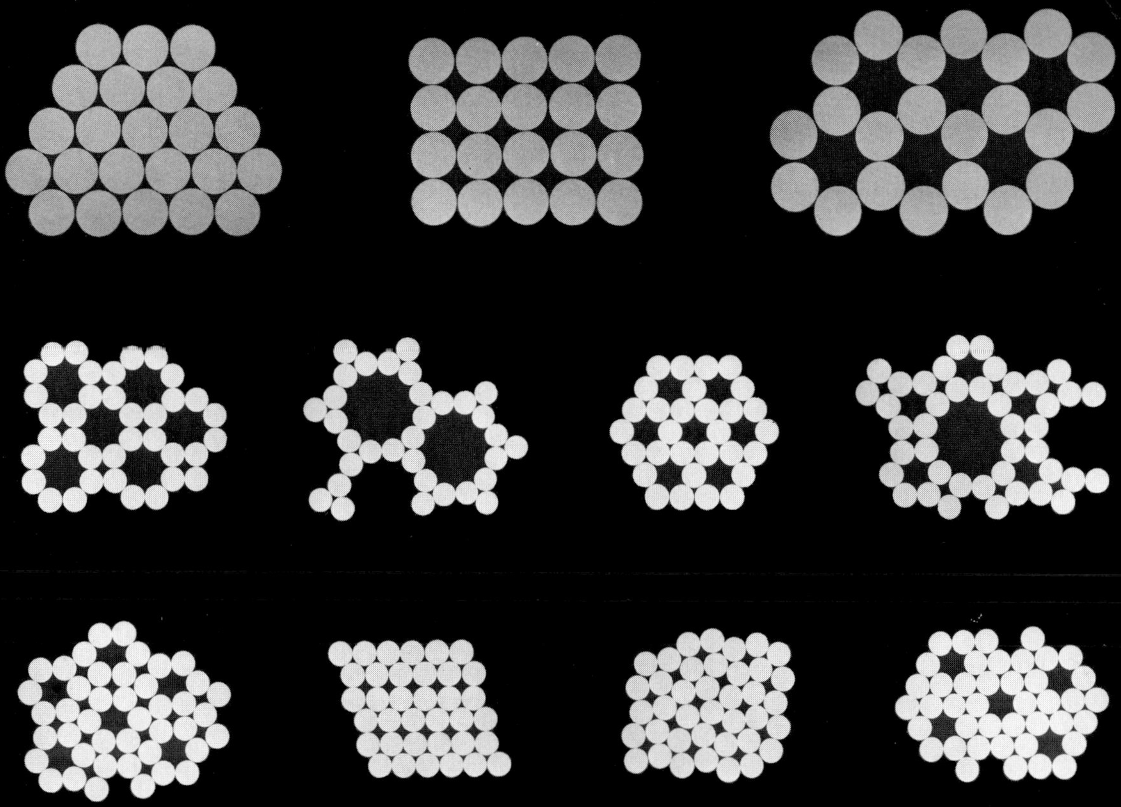

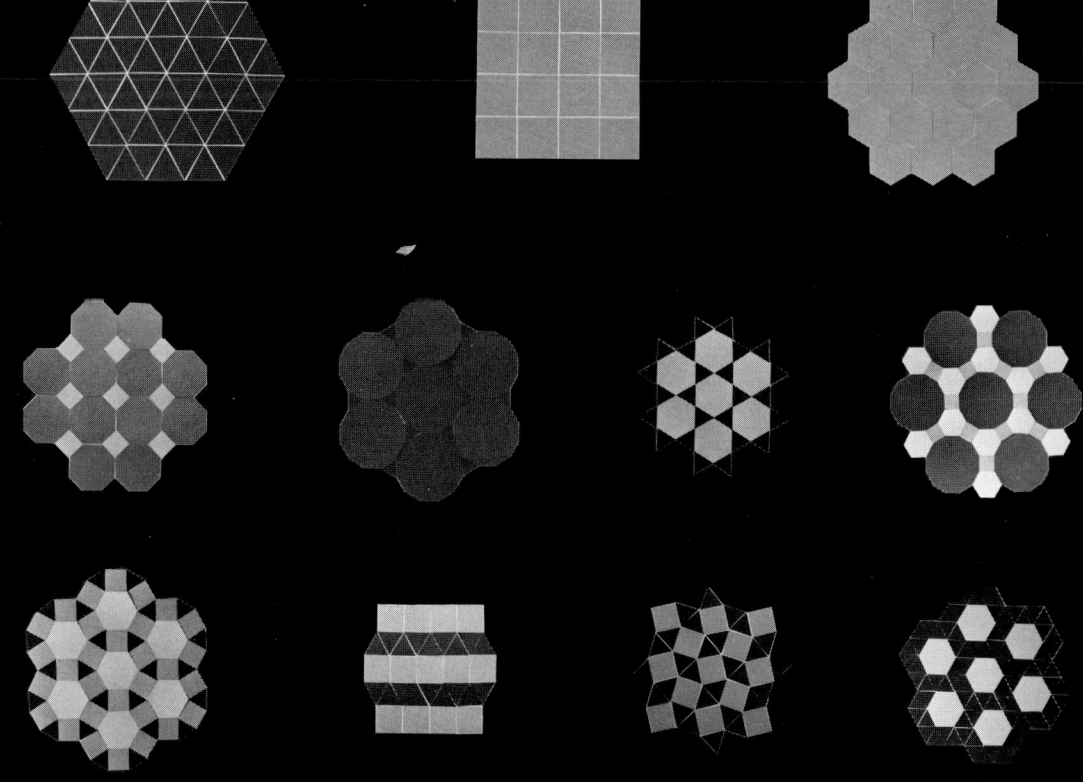

Copper incense burner having 12 spherical pentagonal openworks. Shosoin treasure house, Nara, Japan. Eighth century.

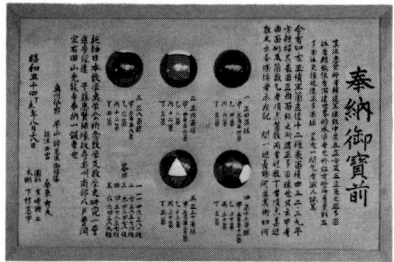

Mathematical problem and its answer on a Japanese traditional votive tablet, a Sangaku. The problem asks for the edge length of each spherical regular polyhedron when the diameter is given. Offered by Hideo Kuwabara in 1979 to Koryuji-temple, Hachinohe, Aomori, to commemorate Eken Shinpo.

together at each vertex. At the extreme left is the regular octahedron having eight regular triangles, four of which are fitted together at each vertex. Right of center is the regular dodecahedron, having 12 regular pentagons, three of which are fitted together at each vertex. On the extreme right is the regular icosahedron, having 20 regular triangles, five of which are fitted together at each vertex.

Each has a circumsphere which touches every vertex, an intersphere which touches the midpoint of each edge, and an insphere which touches the center of each face. The centers of all these spheres coincide with the body center. It is out of this close relationship between regular polyhedra and spheres that regular polyhedral spheres in figure ③ are born. T_6 appears when a regular tetrahedron is projected onto its circumsphere from the body center and all edges are replaced by complete great circles. Similarly, T_8 arises from the simultaneous projection of a cube and a regular octahedron, and T_{10} arises from the projection of a regular dodecahedron and a regular icosahedron.

Conversely, if each is projected onto a suitable plane by parallel lines, such as by the sun's rays, regular polygonal shadows as in figure ⑤ can be cast. Therefore, Critchlow says that shadows of the regular polyhedra are hidden, like shapes of gods, in the regular polygons often seen in arabesques. Figure ⑥ shows some new arabesques designed with the shadows of the regular polyhedra.

A modern mathematician, Klein, proved that every body having rotational symmetry in art and nature is either a regular n-gonal prism C_n, a regular n-gonal pyramid D_n, a regular tetrahedron T, a regular octahedron or cube O, a regular icosahedron or regular dodecahedron I, according to its symmetrical nature with respect to rotation. Regular hexagonal snowflakes and regular pentagonal cherry blossoms have the real and vulgar symmetries D_6 and C_5 respectively.

The temaris shown in figure ② are arranged according to their symmetry groups, ignoring colors. Those shown at the top center of the figure, those second from the right, and those at the right have ideal and sacred symmetries T, O, and I respectively. The groups at the left side and those second from the left belong to C_n and D_n respectively.

Some Japanese traditional mathematicians, the wasan-ka, also knew the mystic beauty of the regular polyhedra, and in particular were aware of the regular dodecahedron and icosahedron. A wasan-ka and priest, called Eken Shinpo, who lived at Hachinohe in the north of Japan at the beginning of the eighteenth century, was proud that he had discovered the regular dodecahedron and icosahedron by himself, and said that both have curious shapes and would be useful for the design of incense burners and medicine chests, and for use by housewives. In particular, a regular dodecahedron can form a die for fortune-telling, using the 12 animals of the Japanese zodiac: rat, cattle, tiger, rabbit, dragon, snake, horse, sheep, monkey, cock, dog, and wild boar. And so he wished to make it widely known.

On the other hand, a contemporary of Eken, wasan-ka and lord Yoriyuki Arima, who lived at Kurume in the south of Japan, said that both polyhedra were widely known and it was not astonishing that someone could make use of them.

In ancient Greece Pythagoras was already aware of the regular polyhedra in connection with regular tessellations. However, he commanded his pupils to keep the existence of the regular dodecahedron secret because of its mysterious property: it is composed of only those regular pentagons that do not appear in his tessellations. Hippasus, notwithstanding, disclosed the secret, and was punished by the gods, drowned in the Mediterranean.

Regular 24-gonal stained glass. Angers Cathedral, France. The upper half, 12 sections, shows the zodiac.

2 The Cosmos

by Plato

A true beauty is born from the ruler and the compass *(Philebos)*. So said Plato, who had inherited the circular and regular polygonal teachings of Pythagoras.

There are set squares, such as the ruler, lying scattered on the earth even now. A set square is a combination of two kinds of right triangles. One is the isosceles right triangle having 45° as the base angle and the other the scalene right triangle having 30° and 60° as base angles. These two kinds of triangles, stoicheia, were the basic elements for Plato. They were superior to the point or circle and were to be found scattered everywhere in the cosmos of that time. As in figure ⑦, six scalene stoicheia construct a regular triangle, and eight isosceles stoicheia construct a square. Thus all of the Pythagorean regular tessellations can be derived from these two kinds of stoicheia. However, the regular tessellations cover over only the two-dimensional plane and are not suitable to describe the three-dimensional space of the cosmos. What forms are suitable for a true three-dimensional object of beauty, drawn with ruler and compass and constructed only of stoicheia?

Plato took many pages of his *Timaeus* to explain how the four regular polyhedra are the shapes of the four elements: earth, water, fire, and air. And not afraid of Heaven's vengeance, he dared to take the fifth regular polyhedron as the outer shape of the cosmos. Therefore, the regular polyhedra soon came to be known as Platonic solids.

Amon Rā as the Holy Ghost of the four elements. Ancient Egypt. According to Champollion.

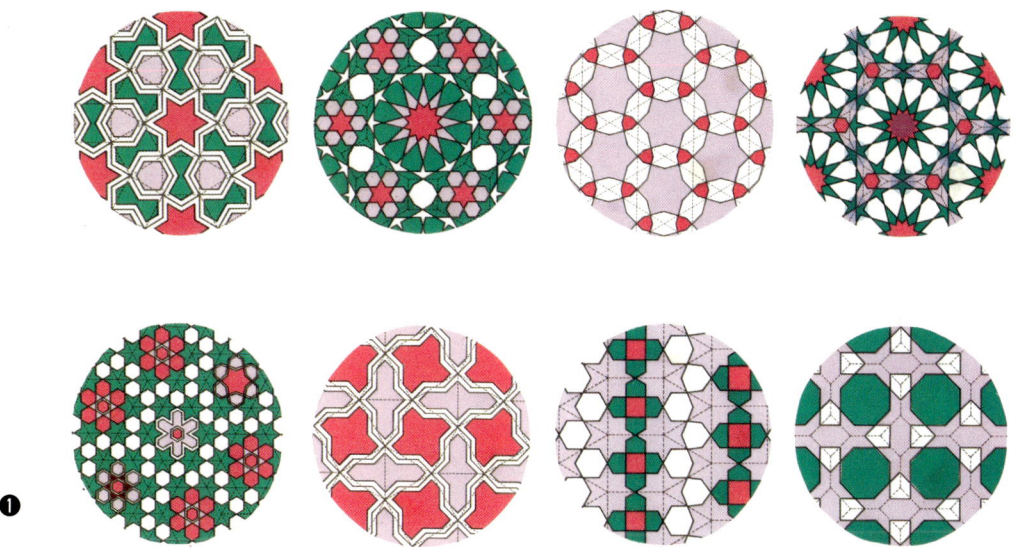

❶

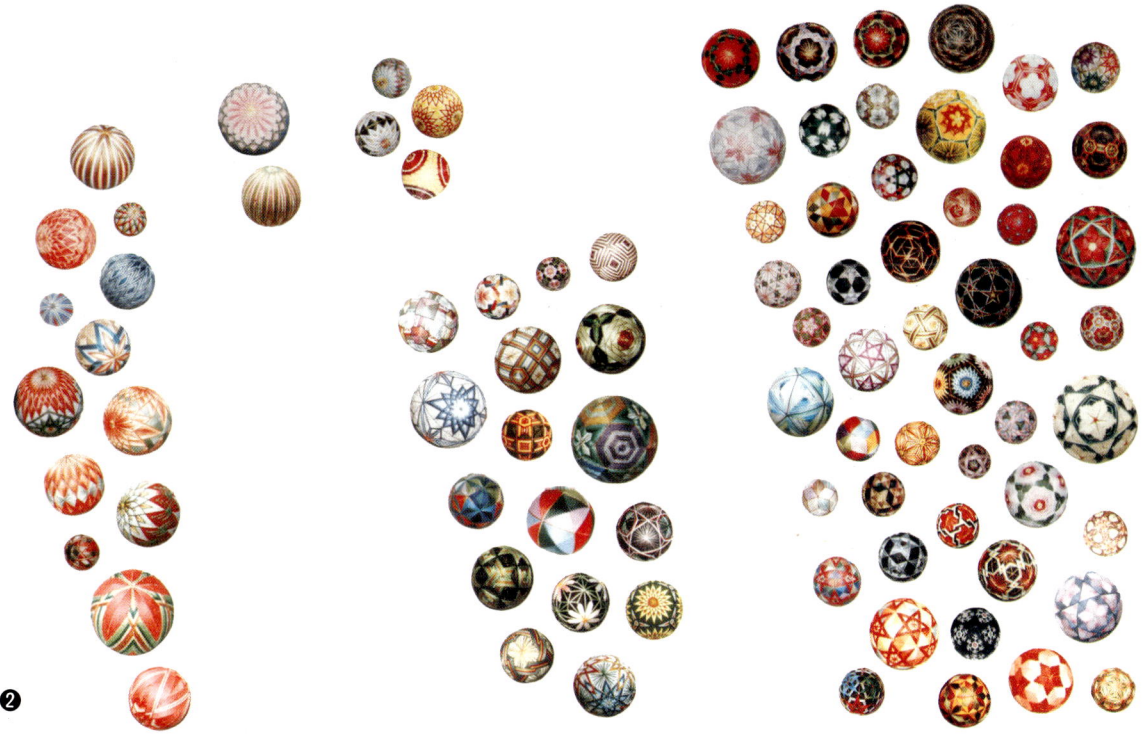

❷

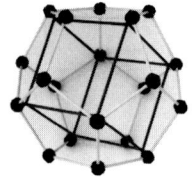

According to Timaeus, an astronomer of the Pythagorean school, the four elements consist of stoicheia, though they are too small to be seen. Fire is the smallest, most pointed, and lightest among the elements because it can easily invade and destroy everything. It is natural, therefore, that the regular tetrahedron composed of four regular triangles or 24 scalene stoicheia like T_6 be taken as the shape of fire, because it is the smallest and most pointed among the regular polyhedra. Water is the largest, smoothest and heaviest because it always flows smoothly into the valleys of the earth. It is natural, therefore, that the regular icosahedron, composed of 20 regular triangles or 120 scalene stoicheia like T_{10}, be taken as its shape.

Air, or vapor, stands between fire and water. It is natural, therefore, that the regular octahedron composed of eight regular triangles or 48 scalene stoicheia like T_8 be taken as its shape. It has the same faces as do "fire" and "water," and it has a number of faces between the numbers of faces of those two.

Fire, water, and air can be mutually transformed: if water is warmed by fire, then air is born, and if air loses fire in the upper atmosphere, then water is born as cold rain or snow.

More precisely, if an element of fire having 24 stoicheia and two elements of air having a total of 96 stoicheia are broken into pieces in the sky, then 120 stoicheia will scatter, constructing an element of water having that number of stoicheia. In other words, the structural formula of water is "Air_2Fire"

Regular polyhedra inscribed in a regular dodecahedron. From top to bottom, a regular tetrahedron and a cube whose vertices coincide with those of the dodecahedron, and a regular octahedron whose vertices are on the midpoints of edges of the dodecahedron. A regular icosahedron, the dual polyhedron to the dodecahedron, is inscribed so that the vertices are on the centers of the faces of the dodecahedron.

Arabesques derived from semiregular tessellations, after Keith Critchlow. The Tetractys is shown by pale blue circles in the right end figure of the top row.

Temaris made by Kiyoko Urata. From the left to the right, those having the cyclic, dihedral, tetrahedral, octahedral, and icosahedral symmetry.

instead of H_2O. Liquor and oil are mixtures of such fire and water, and pepper has a lot of fire.

The remaining element, earth, is as lonely as a stone and as motionless as a mountain, because it has the stable shape of the cube, composed of six squares or 48 isosceles stoicheia like T_8.

The human body is also composed of these elements; when one has an unusually large amount of fire, as a red factor of the blood, one becomes feverish and has a headache because the pointed apices of excess fire invade and poke one's brain. So in order to lower the heat, one uses a water pillow. Then the larger bulk of the element water pushes out the smaller bulk of fire from the body. If there is too much water, one may have loose bowels. It is terrible when there is too much earth because earth is rather binding, and causes one to die and to return to the soil. One who has much fire, though not in poor health, is hot-blooded, and one who has much water is cold-hearted.

In this way, the cosmos is completely filled with stoicheia and with hollow regular polyhedra. Nevertheless, the most secret one, the regular dodecahedron, cannot be seen among them. It is sure that Plato was troubled and looked devoutly at the deep starry sky, as everyone does. At such a moment,

◄ ③

Basic patterns of temaris. From left to right, "octasect" pattern T_8, "hexasect" pattern T_6, and "decasect" pattern T_{10}.

◄ ④

Regular polyhedra. From left to right, a regular octahedron, cube, regular tetrahedron, regular dodecahedron, and regular icosahedron.

◄ ⑤

Regular polygonal shadows of regular polyhedra. From left to right, shadows of regular octahedra, cubes, regular tetrahedra, regular dodecahedra, and regular icosahedra.

◄ ⑥

Original arabesques using regular polygonal shadows of regular polyhedra.

Development of a regular dodecahedron with symbols of the Zodiac.

he understood that the outer shape of the cosmos, a natural vessel for the four elements, would be exactly a regular dodecahedron. Various celestial bodies were attached to its interior. A philosopher of the Middle Ages, Bacon, wrote in *Opus Majus* that Plato was correct because the other four regular polyhedra can be easily inscribed in a regular dodecahedron, placing the former's vertices on the latter's vertices, at midpoints of edges, or at centers of faces.

In Kepler's opinion, the zodiac is separated into 12 because the cosmos has a regular dodecahedral outer shape (*Harmonices Mundi*). Socrates, too, proposed a dodecahedral cosmos, just before his death. His cosmos is seen from the outside as if it were a temari (ball) painted in brilliant colors (*Phaidon*).

Figure ⑧ shows Plato's cosmos represented as a temari drifting among scattered stoicheia. The face of the temari is decorated with strings, which divide each regular pentagon into particular stoicheialike triangles. The famous symbols of the Pythagorean school, the pentagrams, are also seen. This pattern can be obtained when T_6, T_8, and T_{10}, the basic shapes of four elements, overlap. Four colors—white, black, red, and yellow—were thought to be the primary colors in ancient Greece, because they were those most easily compounded at that time. The atmosphere around the earth was not blue.

One of regular dodecahedral dice with symbols of the Japanese Zodiac. An artifact of Takayama, Gifu.

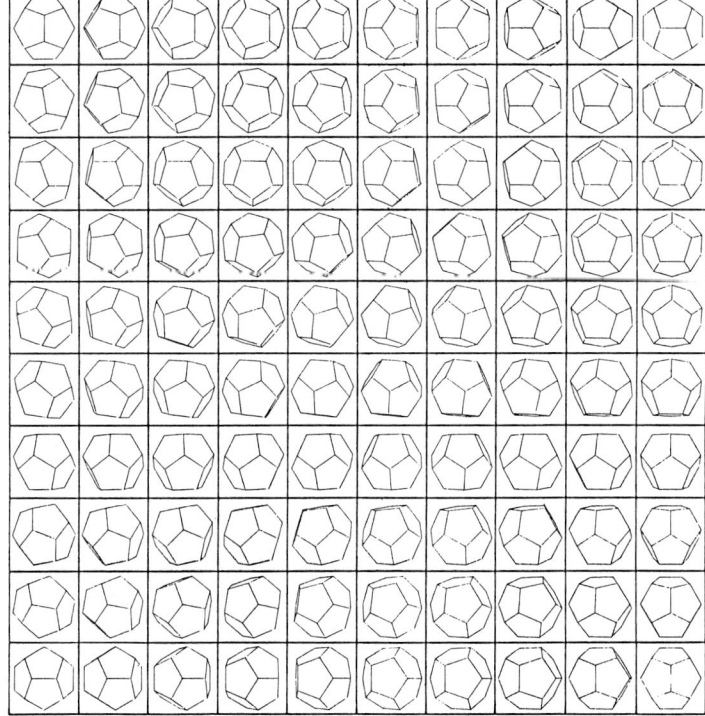

Various projections of a regular dodecahedron which can be seen as a combination of stereograms, using crossed eyes. Computer graphics by Taro Tajima and Kunio Kondo.

⑦ ▶

Stoicheia.

⑧ ▶

Plato's cosmos as a temari in scattered stoicheia.

3 Nature

by Aristotle

Duality of regular polyhedra. From left to right, cubes and regular octahedra, two regular tetrahedra, regular dodecahedra, and regular icosahedra.

◀ ⑩

Archimedean solids and their regular polygonal shadows. The figure shown in the central column is a truncated tetrahedron. Those of the left column are of the Cube family: from top to bottom, a truncated cube, truncated octahedron, cuboctahedron, truncated cuboctahedron (great rhombicuboctahedron), rhombicuboctahedron (small rhombicuboctahedron), and snub cuboctahedron. Those of the right column are of the Dodecahedron family: from top to bottom, a truncated dodecahedron, truncated icosahedron, icosidodecahedron, truncated icosidodecahedron (great rhombicosidodecahedron), rhombicosidodecahedron (small rhombicosidodecahedron), and snub rhombicosidodecahedron.

Japanese cake box having a rhombicuboctahedral shape (left) and its pseudorhombicuboctahedral appearance, a mistaken assemblage.

According to Plato, all of the cosmos is composed of regular polygons and polyhedra. Many persons, especially doctors and biologists, have been revolted by Plato's theory of the polyhedral constitution of nature. Wouldn't our bodies be injured if our blood were made up of sharp-pointed tetrahedra?

Aristotle, a great biologist, fathered by a doctor, and the prize pupil of Plato, was one of them. He was distressed at the outlandish theories of his respectable teacher. For a time, Aristotle seemed to lack eloquence on the subject. But when the occasion was offered by the death of his teacher, he published his own book on astronomy, *De Caelo*, opposing *Timaeus*. In it, he had marginal agreement with Plato on the ideas that the fundamental figures of the cosmos are circles and spheres, that there are four elements in nature, and that all must be the work of the Creator. No one in those days could oppose such matters of common knowledge. Aristotle said it is absurd to think that men have headaches because the pointed apices of regular tetrahedra invade and poke their brains. It would make us rather crazy. Plato said that the four elements are constructed by meeting and parting of two-dimensional facets, stoicheia. If so, what is in the enclosed spaces of the elements? These inner spaces as "eidos" (three-dimensional forms) are the elementary factors to form the cosmos. The cosmos having only the two-dimensional stoicheia is nothing but a sham. Plato assigned the regular te-

Regular tetrahedral water chestnuts, which are said to have been scattered on the floor as peculiar weapons by a Ninjya, a Japanese traditional spy. Their pointed apices injured the soles of feet of his pursuers.

trahedra to fire because of their smallness, but there are also larger polyhedra. Could a regular tetrahedron larger than a regular icosahedron be used to cool eatables, like the refrigerator is used in the twentieth century? It was explained that fire is hot because of the pointed apices of tetrahedra, but regular icosahedra also have pointed, though less sharp, apices. Is a triangle hot when drawn on a papyrus? If so, a fire will occur. It is said that the element earth has the shape of a cube for stability, but the other elements also become stable when they find their own "topos." A regular icosahedron cannot be cut in such a way that all of the pieces form icosahedra, and however many icosahedra we may use, we cannot closely surround a central one. Therefore, the element water would not behave like the water in the sea, if Plato was correct.

Aristotle attacked Plato, making use of the legend that Pythagoras was aware of only two kinds of space fillings by regular polyhedra in addition to his tessellations: one is the polycube, which is composed only of cubes, and the other the octet honeycomb, which is composed of regular tetrahedra and octahedra. This author prefers to think that the word "pyramids" in *De Caelo* means both the regular tetrahedra and octahedra. There are other opinions that "pyramids" means the tetrahedra only, and Aristotle mistakenly thought that regular tetrahedra can fill 3-space. In response to this riddle,

Octet-truss as an unusual spatial kite. Designed by Alexander Graham Bell.

Cubic lattice as a usual jungle gym.

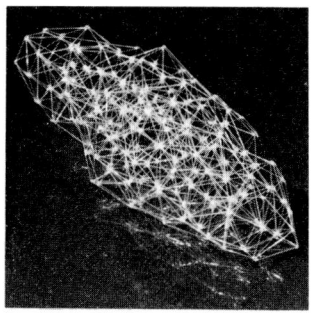

Glass-work Galaxy, *a liquid in the universe, which looks like a molecular structure of the water. By Yumiko Yoshimoto.*

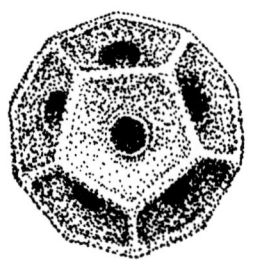

Regular dodecahedral pollen of Gypsophila elegans.

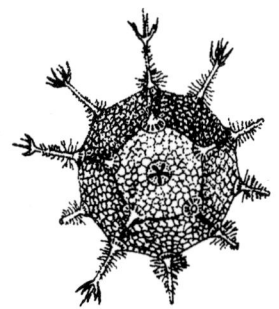

Regular dodecahedral skeleton of a radiolaria. According to Ernst Haeckel.

7 ▶

Models of viruses with photos of DNA in viruses as background. From top to bottom, Herpesviruses, Rice dwarf viruses, Papovaviruses, and Adenoviruses.

Bacon minutely examined spatial arrangements of regular triangles and of squares relating to the above mentioned two space fillings in *Opus Majus*.

Aristotle says that if the cosmos were composed of regular polyhedra, there would therefore be only two variations of eidos in Plato's completely filled cosmos.

Aristotle's understandable disobedience goes on still further. Following this original controversy, it is said that all men are pupils either of Plato or of Aristotle. Such a man of absolutism as Newton is a pupil of Plato, whereas such a man of relativity as Einstein is a pupil of Aristotle. Regrettably, today Aristotle is more respected as the father of natural science.

However, the world on an atomic scale reminds us once more of Plato. For example, according to Bernal, a contemporary physicist, the molecular structures of liquids are derived from polyhedra whose faces are regular triangles and which have icosahedral symmetry, such as regular icosahedra. Of course, 3-space cannot be filled stably with such polyhedra, and it is thus that the molecules of liquids staggeringly move and flow, trying to become stable.

The territory of viruses is also inhabited by regular icosahedra. Figure 7 represents the models of various viruses, with photos of DNA in viruses for a background. The upper right figure represents periodic arrangements of viruses following the Pythagorean tessellations.

Plato said that when there are too many regular icosahedra in our body, we get sick. Even if this is not so, regular icosahedra are dangerous.

Pollen, which floats in the air or is carried by the current, has a larger bulk than viruses. For example, *Gypsophila elegans* has the regular dodecahedral, *Oenothera odorata* the regular tetrahedral, and *dalia* the cubic shape. Most of them are spherical in the water, having orifices in regular position. Fejes-Tóth quotes a biologist to the effect that the number of equiradial and mutally contacting circles is maximum when they are drawn on the surface of a pollen with each orifice as the center. Therefore, in the case of 12 or 20 orifices, they are arranged at the center of the faces of regular dodecahedral or icosahedral spheres respectively.

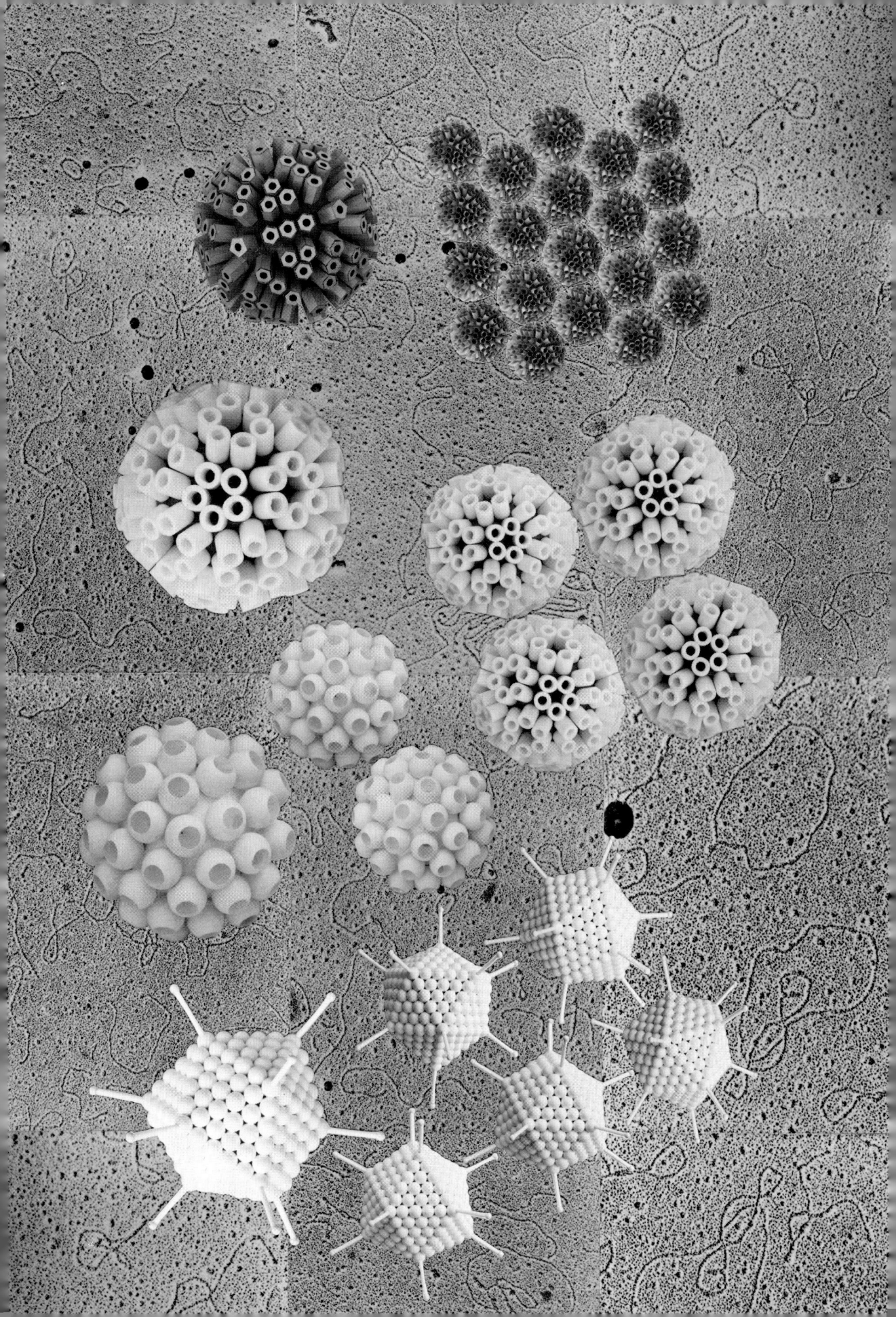

Crystals of pyrite: (Upper) *cubic and* (lower) *regular dodecahedral.*

The regular polyhedra appear as the shapes of skeletons of radiolaria in the sea. E. H. Haeckel depicted them in great detail, as is well known. Viruses, pollen, and radiolaria preserve their polyhedral forms while alive and floating in the air or in the sea. By comparison, recall that minerals, which have been lying lifeless under the ground since Plato's time, crystallize in the shapes of regular polyhedra. Figure 8 shows polyhedra as crystals of natural diamonds. All have shapes derived by truncating cubes and regular octahedra. Among minerals, pyrite is mined in every regular polyhedral form except as a regular tetrahedron. Dodecahedral crystals of pyrite exist, and may have been the inspiration for the dodecahedron of Etruscan origin, found at the foot of Mt Loffa, in Italy, in 1885. T. L. Heath, a modern historian of mathematics, holds the opinion that Pythagoras may well have been aware of the regular dodecahedron, though he kept the secret.

Such regular polyhedra have remained in the brains of many scientists, not as evil viruses but as accurate knowledge, ever since Euclid's *Stoicheia* was written, in the third century B.C.

Many superior geometrical brains have pondered the famous postulate on parallelism, described at the beginning of Book 1 of the 13 books of the *Stoicheia*. It was Proclos, a philosopher of the fifth century, who concentrated his and others' attention to the end of Book 13, because it is there that the existence of the five, and only five, polyhedra is proven. In fact all 13 books are like a Bible, intended to explain regular polyhedra. Regular polyhedra have played a leading role in geometrical and mathematical knowledge in science since the time of Euclid.

Aristotle was a pre-Euclidean man.

Models of natural diamonds born between a cube and regular octahedron.

The Family by Archimedes   4

There is the earth in contrast to the heaven, the square to the circle, and Aristotle to Plato. All come in pairs.

Mutually dual polyhedra also come in pairs. Two polyhedra are called mutually dual when the vertices of one can be replaced by the faces of the other, and vice versa.

Among regular polyhedra, a tetrahedron is dual to itself, a cube is dual to an octahedron, and a dodecahedron to an icosahedron, as shown in figure ⑨. Mutually dual polyhedra resemble each other in that they have the same number of edges, and they have the same symmetry. The three basic patterns of temaris in figure ③ are erected from pairs of mutually dual regular polyhedra. Kepler made a distinction between the Cube family, whose father is a cube and mother a regular octahedron, and the Dodecahedron family, whose father is a regular dodecahedron and mother a regular icosahedron. A regular tetrahedron is single (*Harmonices Mundi*).

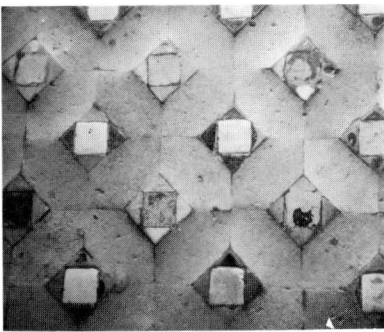

Arabesque on the floor as shadows of truncated octahedra and cuboctahedra. Alhambra.

The Cube and Dodecahedron families each have six children, and the unmarried regular tetrahedron also has a child. It was Archimedes who bathed these 13 children and brought them up. They dressed in two or more different kinds of regular polygons, fitted together similarly at each vertex, as in figure ⑩. Their regular polygonal shadows are also shown in gray.

Each of the six at the left is a child of the Cube family and each at the right, the Dodecahedron family. The one in the center is a child of the regular tetrahedron.

One can learn from the names of these polyhedra the environment in which they were born. For example, a truncated cube is derived by the truncation of the vertices of a cube. A cuboctahedron is a half-breed of a cube and a regular octahedron. A rhombicuboctahedron (small rhombicuboctahedron) has more squares—that is, rhombi—than a cuboctahedron. A snub cuboctahedron is derived from the distortion of all faces of a cuboctahedron. A truncated cuboctahedron (great rhombicuboctahedron) is obtained when the rectangles which appear as the regions around vertices of a cuboctahedron are replaced by squares. Similar constructions can be carried out within the Dodecahedron family.

Archimedean hexagonal prism (upper) *and anti prism* (lower).

According to the inquiry by Heath, Heron in the first century B.C. said these 13 polyhedra had been found by Archimedes. Pappus in the third century reexamined them. Kepler in the seventeenth century was the first to draw all the pictures.

Thus many geniuses have taken an interest in the 13 solids, but they have come to be called the Archimedean solids despite the fact that the manuscript written by Archimedes was lost in the burning of Alexandria.

Curiously enough, a pseudo-rhombicuboctahedron was recently found in different parts of the world at almost the same time. The shape is obtained when a caplike part having four triangles and five squares of a rhombicuboctahedron is rotated 45° around the axis of symmetry which passes through the centers of the parallel square faces. Despite the rotation, the facial and vertical regularity of the Archimedean solids is not altered.

In Japan, Hiraku Toyama, a modern mathematician, also discovered this new polyhedron when his grandchild mistakenly closed the cap of a Japanese rhombicuboctahedral cake box, seen in figure ⑪. When this pseudo-rhombicuboctahedron is added, the number of the Archimedean solids becomes 14. Kepler also said, perhaps mistakenly, that there were 14 *(On the Six-Cornered Snowflake)*.

In addition, there are innumerable Archimedean prisms and anti-prisms also having the same regularity as the Archimedean solids. The upper and lower bases of both are arbitrary congruent regular polygons, but the lateral faces of prisms are squares and those of anti-prisms are regular triangles. Therefore, all 13 Archimedean solids, a pseudo-rhombicuboctahedron, infinitely many Archimedean prisms and Archimedean anti-prisms are all called semiregular polyhedra today.

Of these, only the original 13 Archimedean solids, however, can be inscribed in a regular tetrahedron so that an appropriate set of four faces can overlap each other. It is not strange, then, that Archimedean solids, like angels, be put in sacred positions next to the Platonic solids, their gods. Some people conclude that Christ and the Apostles cast their shadows on the 13 solids. A truncated tetrahedron born of an unmarried regular tetrahedron represents Christ. In *The Last Supper* by Dali, Christ and the Apostles are arranged under a regular dodecahedral dome. The Jusan-butsu, a set of 13 Buddhas which has been worshiped in Japan since about the thirteenth century, may have the same origin.

The dual polyhedra of the Archimedean solids are called Archimedean duals or Catalan's solids. Unlike the Platonic solids, Archimedean duals differ entirely from the Archimedean solids. The 24-, 48-, and 120-hedron, derived from the basic patterns of temaris shown in figure ③, are dual to a truncated octahedron, a truncated cuboctahedron, and a truncated icosidodecahedron respectively.

In contrast to the regular polyhedra and semiregular polyhedra, which have been known from ancient times, convex polyhedra, composed of regular polygons but lacking regularity around vertices, were found only in the twentieth century. Those of them which have only regular triangular faces are the eight deltahedra in figure 9. They are 4-, 6-, 8-, 10-, 12-, 14-, 16-, and 20-hedra. There is no 18-hedron.

If not convex, there are innumerable polyhedra having only regular triangles as faces. For example, figure 10 shows star-shaped solids whose original pictures were drawn by Leonardo da Vinci for Pacioli's *De Divina Proportione*. If Plato, who always admired the heavens, had been aware of these, he would have thought that they would be new elements transported from an unknown star.

Planar stone monument, an Itabi, for Japanese Buddhists, on which the names of 13 Buddhas are inscribed in Sanskrit. The largest uppermost circle shows Vairocana, the sun, and 12 small circles below it, which are separated into two columns, mean various Buddhas. The stone statue is Kukai. Kiyotanidera temple, Tokyo. Seventeenth century.

Luca Pacioli (center) and a transparent rhombicuboctahedron.

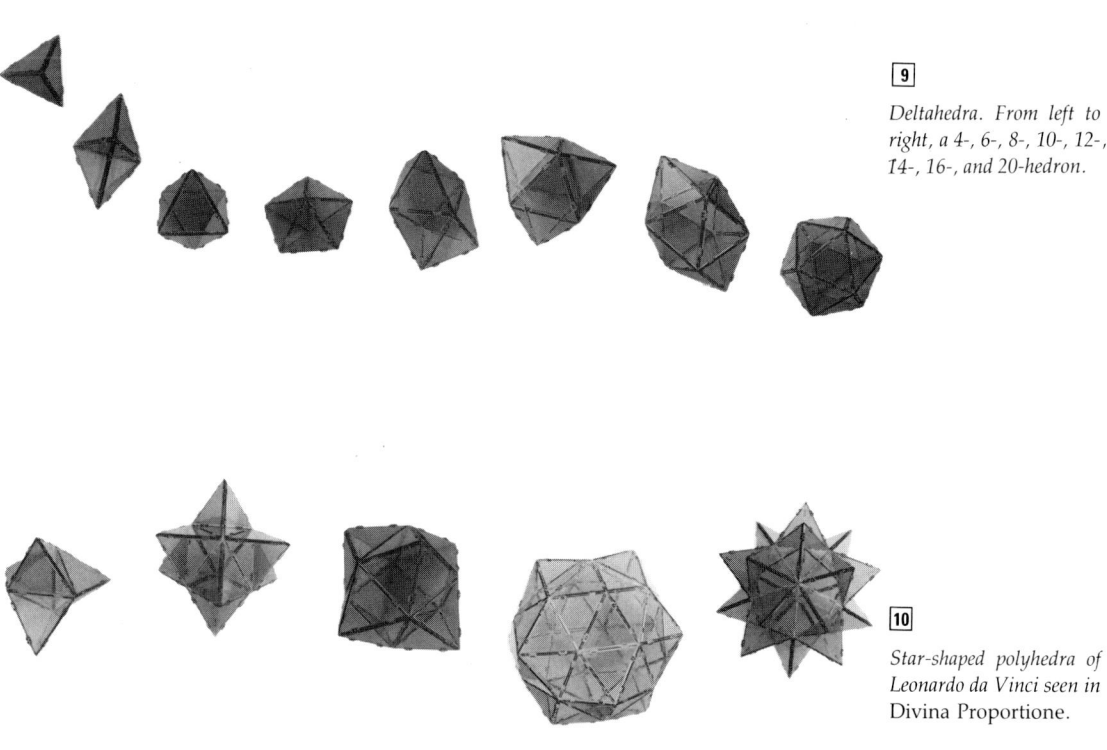

9

Deltahedra. From left to right, a 4-, 6-, 8-, 10-, 12-, 14-, 16-, and 20-hedron.

10

Star-shaped polyhedra of Leonardo da Vinci seen in Divina Proportione.

Regular icosahedral (center) *and cuboctahedral* (both sides) *dice. British Museum. First century.*

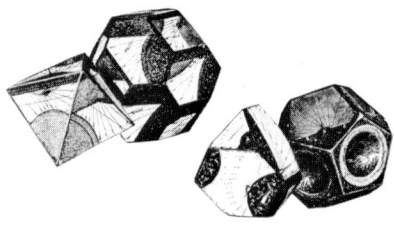

Polyhedral sundials of Italy. Sixteenth century.

By contrast, the only convex polyhedron having just square faces is the cube, and the only one having regular pentagons is the regular dodecahedron. There are no convex polyhedra having only regular hexagons or more-sided regular polygons.

What shapes are possible for convex polyhedra having arbitrary regular polygonal faces, and with no restriction on the placement of the faces around the vertices? According to V. A. Zalgaller and N. W. Johnson, there are 92 kinds with the exception of regular polyhedra, Archimedean solids, and Archimedean prisms and anti-prisms. Figure ⑫ shows pairs of front and rear views of each, except the deltahedra and the pseudo-rhombicuboctahedron. All are classified according to their symmetry. Groups on the extreme left, those third from the right, and the rightmost belong to the Cube family, and the rest belong to the Dodecahedron family. There are a half dozen pairs that appear identical but differ from each other in the same way that a pseudo-rhombicuboctahedron differs from a rhombicuboctahedron.

The four elements to which Plato gave the shapes of regular polyhedra were also widely believed to possess magical properties, until Boyle in the seventeenth century refuted medieval alchemy. Even in Goethe's *Faust,* written at the beginning of the nineteenth century, the agonies of the dying elements are found.

With Boyle began the concept of the modern elements, including hydrogen as the atomic number 1, helium as 2, and so on. Today, 92 naturally occurring elements, ending in uranium, are known. If artificial ones are added, there are more than 100 elements.

The 92 elements found in nature may have the shapes of 92 convex polyhedra with regular faces, though many have irregular shapes. According to physicists today, nature has rather irregular features, which do not preserve the parity between right and left.

In the 1930s, however, a crisis occurred: the proton, electron, and neutron were discovered to be the only three elementary particles in nature. They are fewer in number than the four elements. These three may be formed from the regular tetrahedron, cube, and regular dodecahedron, of which Kepler took particular notice, since they have three different types of faces. They can also be formed from the regular tetrahedron, octahedron, and icosahedron, which have been given new meaning by Fuller, as will be mentioned later.

The number of known elementary particles, however, has been increasing year by year. Recently, more than 150 were known. If they are enumerated in a different way, they number about 400. The situation is rather complicated.

A more basic particle, the quark, which has recently been devised, is useful for simplification, because many elementary particles are considered to be formed of only this quark. But today about 20 sorts of quarks are found.

After all, how many elements are there in nature? Eight deltahedra? Thirteen or fourteen Archimedean solids? Probably not any of these. It should rather be six, which is the number of four-dimensional regular polytopes, or perhaps three, the number of five- or higher-dimensional regular polytopes, as will be shown later. Six "flavors" making up the fundamental nature of quarks have already been devised, and it is known that each flavor has three "colors."

Convex polyhedra having regular polygonal faces. Front and rear views of 86 kinds are shown by pairs. The other six kinds are of deltahedra and a pseudo-rhombicuboctahedron.

Kepler's arrangement of the circular orbits of the planets following designs of regular polygons.

⑭ ▶

Reconstruction of Kepler's cosmic cup.

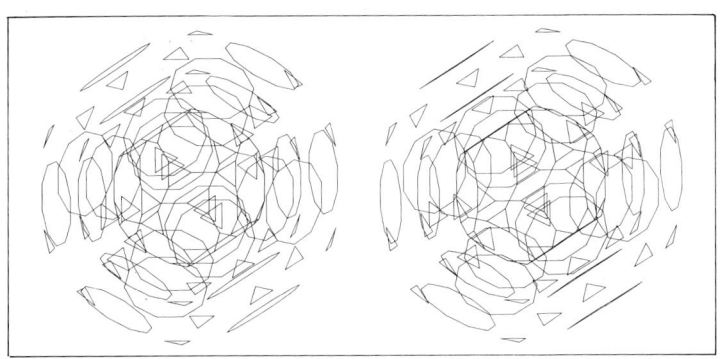

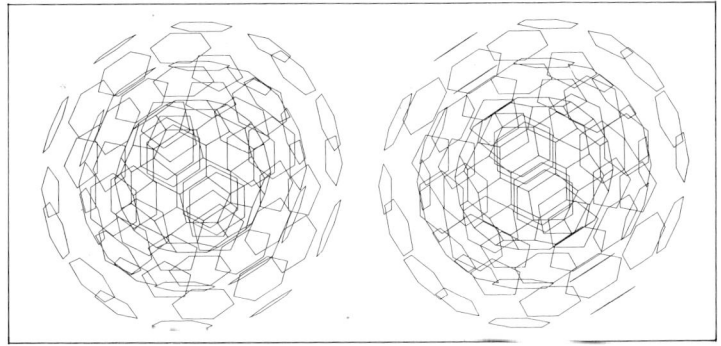

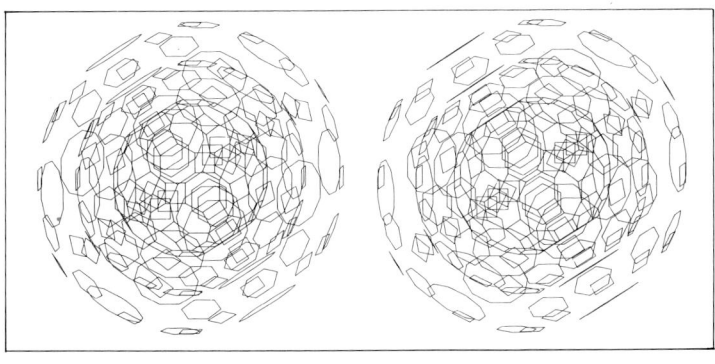

Explosion of the Archimedean solids. From top to bottom, a truncated dodecahedron, truncated icosahedron, and truncated icosidodecahedron. Stereograms by crossed-eyes. Computer graphics by Taro Tajima and Kunio Kondo.

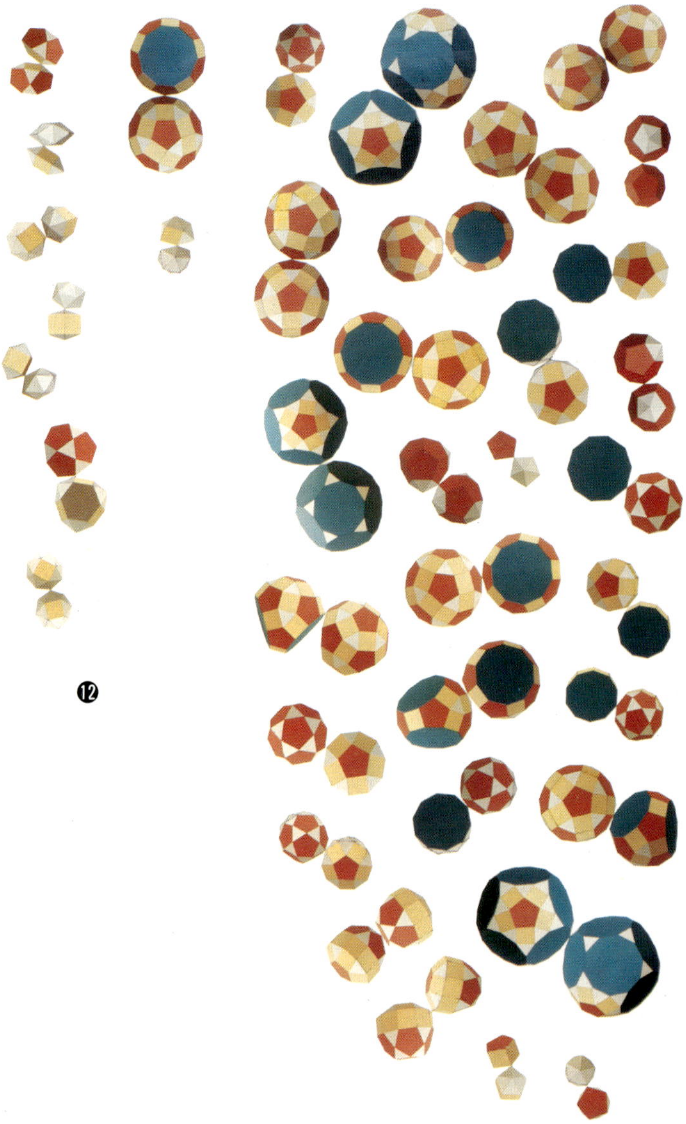

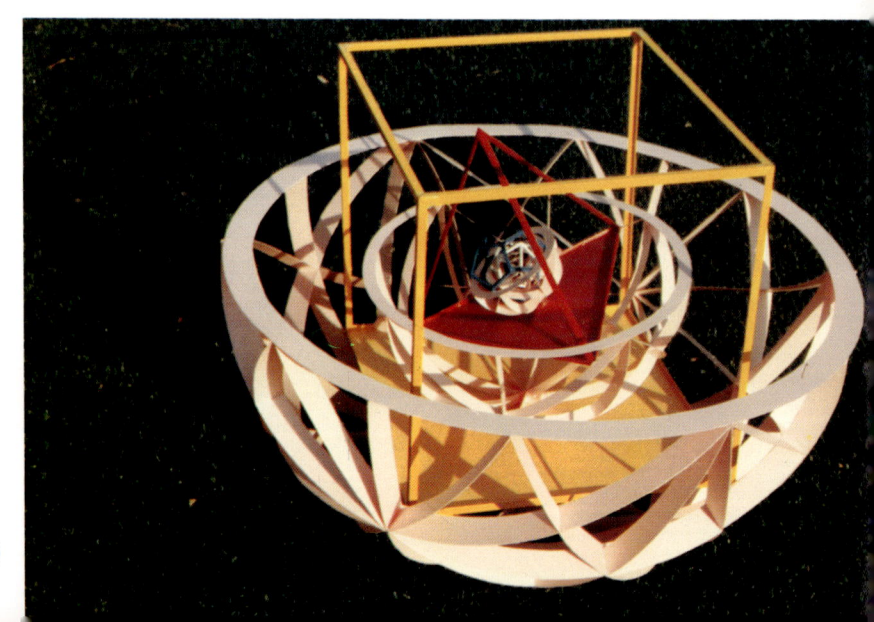

❷

❸

❹

⓯

5 The Dream by Kepler

God as an architect of the universe. According to a bible of the thirteenth century.

Fundamental patterns of geodesic spheres. The alternate subdivision (right), all of whose edges are perpendicular to a spherical icosahedron; the triacon subdivision (center), all of whose edges are perpendicular to a spherical icosahedron; and their combination as a temari (left).

According to Plato, polyhedral elements of nature, which are too small to be seen, are contained in a polyhedral vessel, the universe, which is too large to be observed.

As an astrologer, Kepler was convinced of Plato's geocentric system; as an astronomer, he was impressed by the Copernican heliocentric system. He designed a "cosmic cup," following the ideas of Plato and Copernicus.

The environment at the dawn of the modern age in which Kepler was born and raised was as dark as during the Middle Ages. Thus he had a craze for astrology and became so familiar with the heavens as to openly defend Copernicus, though those who doubted the geocentric system were thought to be faithless people or lunatics to be burned at the stake.

Kepler's natural sagacity and eagerness fitted him for mathematics, and he chose as his profession to be a private lecturer in mathematics, with the intention of becoming a theologian. This was the beginning of his wanderings between the Middle and Modern Ages, between God and man, religion and science, between astrology and astronomy, hesitatingly, erringly, and reflectingly.

One day when he was 25 years old, he was absorbed in drawing a beautiful astrological figure on a blackboard and was thinking as an astronomer about why the six planets move on their circular orbits, with radii determined by the observed values of Copernicus. The figure was a large circle inscribing a regular triangle, which in turn inscribed a small circle. Having finished the drawing, he was startled, for he was convinced that he would be able to solve a puzzle about the arrangement of the orbits. He announced the conclusion of his lecture to his few students, and in a hidden place drew a figure such as figure ⑬.

First a large circle for the orbit of Saturn was drawn and a regular triangle as the first regular polygon was inscribed in it. Then its inscribed circle became the orbit of Jupiter. Similarly, he made Mars, Earth, Venus, and Mercury move on the inscribed circles of a square, regular pentagon, regular hexagon, and regular heptagon, respectively.

Johannes Kepler.

Kepler, who excitedly compared the result with the values of Copernicus, was startled again. They are not the same. However, he thought that he came nearer to the truth because of the mistake. It was wrong to consider plane polygons. Polyhedra in 3-space should rather be applied to obtain the spatial orbits. Which polyhedra are suitable for the divine movement of the planets? They should be the regular polyhedra.

First, a large sphere whose geodesic was the orbit of Saturn occurred to him. A cube inscribed in it was Plato's Earth, and Jupiter seemed to go around on a geodesic of the inscribed sphere. Similarly, Mars moved on a geodesic of the inscribed sphere of a regular tetrahedron, Plato's fire. The Earth moved on the inscribed sphere of a regular dodecahedron, the Platonic outer shape of the cosmos. Venus moved around a regular icosahedron symbolized by Aphrodite born from the foam of the sea, thus from water. Mercury traveled around a regular octahedron, suggesting its airlike movement. After a careful examination, the values for all the planets except Saturn and Mercury almost agreed with the values observed by Copernicus. As for Saturn and Mercury, Kepler thought Copernicus had been mistaken because of their too great distances from the Earth. However, worried about these discrepancies, he gave some thickness to each sphere to absorb accidental errors and to include the orbits of satellites such as the moon. Thanks to this correction, Kepler was later able to state his law that "each planet goes on the elliptical orbit" within the limit of the thickness of each sphere. Thus mankind went beyond the circles of Pythagoras, in a way which later helped us to land on the moon.

This surprising discovery of elliptical orbits was recorded at once in a book entitled *Mysterium Cosmographicum*, and a petition to construct a model "cosmic cup" was presented by Kepler to his patron. According to Kepler's plan, the whole cosmos is represented as in a hemispherical cup having the diameter of more than 1 meter, on whose geodesic Saturn moves. All was to have been made in silver and various jewels were to be set thereon: a diamond for Saturn, a ruby for Ju-

Star-shaped model of the solar system used by Gauss in his lecture. Apices of thorns determine the orbits of the planets.

Plan of the universe explained in Kosá-Sāstra scripture.

Gorinto (solid at the center) and three Stupas colored in a traditional manner.

Regular polyhedra Mandala.

⓰

⓱

⓲

⓳

⓴

11
Geodesic spheres floating high in the sky. The second one from left shows the inner scene.

◀ ⑲
Appearance of the microscopic world as explained in Kosá-Sástra scripture.

◀ ⑳
Packing of Fuller's universe by spherical convex cuboctahedra, concave octahedra, and concave cuboctahedra.

piter, a pearl for the moon, and so on. Furthermore, the cup was to be filled with suitable drinks: special whisky for the Sun, wine for Saturn, brandy for Mercury, and so on.

Kepler persuaded the patron to make the cup, showing him the blueprint. Owing to its great cost, however, the good-natured patron hesitated, and the project was never realized.

Kepler might have dreamed of such a model as in figure ⑭. The hemisphere for Saturn, having a diameter greater than 1 meter, is decorated by the pattern of T_{10} of figure ③, made of 15 geodesic lines (great circles). Jupiter and Mars are on the hemispheres having the patterns of T_8 made of 9 geodesic lines, and the Earth, Venus, and Mercury are on those having the patterns of T_6 made of 6 geodesic lines.

By the cosmic cup it was revealed to Kepler that regular polyhedra act as hooks, so that the heavens do not fall to the ground.

About 20 years later, he wrote *Harmonices Mundi* in spite of the great unhappiness and misfortunes he met in the world: The Thirty Year's War, death in his family, and the task of choosing a new wife from among 11 applicants.

In this book, he says that regular polygons which can be drawn by a ruler and a compass and have ability to fill a plane or a spherical surface can do much for the harmony of the world.

Further, deeply impressed by the existence of various simple ratios between the numbers of vertices, edges and faces of each regular polyhedron—such as 2 : 3 between the numbers of vertices and edges of a regular tetrahedron—Kepler described in the book the famous third law having a simple ratio: the second power of periods of revolution of any two planets are proportional to the third power of their mean distances from the sun.

Newton explored Kepler's thorny polyhedral world and at last demonstrated mathematically Kepler's third law.

Later, Gauss, a mathematician and astronomer, showed a thorny star-shaped polyhedron at his lecture. It is said that he also showed a model of Kepler's cosmic cup, with Kepler's words: the cosmos can be expressed by elementary geometrical figures.

In contrast to brilliant macrocosmic discoveries, Kepler's mediocosmic life became gloomy, and in loneliness he died on a journey, leaving a scientific romance, *Dream*, in which a journey to the moon is undertaken.

He was unable to dream, however, that at the end of the eighteenth century a law on the arrangement of the planets having a simpler ratio than his third law would be discovered by Bode and Gauss: the mean distance of each planet from the sun is proportional to the value $0.4 + 0.3 \times 2^n$, where $n = -\infty$ in the case of Mercury, $n = 0$ for Venus, $n = 1$ for the earth, $n = 2$ for Mars, $n = 4$ for Jupiter, $n = 5$ for Saturn. Later, an asteroid Ceres and the planet Uranus were discovered, fitting $n = 3$ and 6 respectively; both were unknown to Kepler. Further, he could not foresee that in 1959 a regular dodecahedral insignia of the Soviet Union would be put on the moon, following one of his own elliptical orbits.

Bode's law has larger aberrations as the scale of the universe (n in the law) becomes larger. One man who was convinced that he could explain without aberration the end of the universe when $n = \infty$ is Richard Buckminster Fuller, an architect and scholar of the universe. Koestler said in the 1950s that Plato's *Timaeus* and Kepler's *Harmonices Mundi* are the two and only two synthetic constructions of the universe. In 1975, however, a third one appeared in Fuller's *Synergetics*. *Synergetics 2* was subsequently published in 1979.

Is it coincidence that all of them come to the conclusion that polyhedra are the very ark of the universe? Or is it a fact that the universe is filled with polyhedra?

The word "synergetics" as Fuller used it was born in opposition to traditional mathematics, from a technical term "synergy" occasionally used in biology and chemistry. Microscopic molecules or cells in nature, when they act in concert, exhibit an extraordinary power which is difficult to predict. For example, though composed of similar molecules or cells, a green caterpillar becomes a butterfly, not a chicken,

Fuller's geodesic dome as the Biosphere. Montreal. Directly after the fire.

Fuller holding a combination of Yoshimoto cubes shown in thirty-seven (left), Yoshimoto himself (right), and Itsuo Sakane (center). In Tokyo, 1982: shortly before Fuller's death.

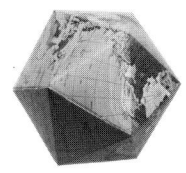

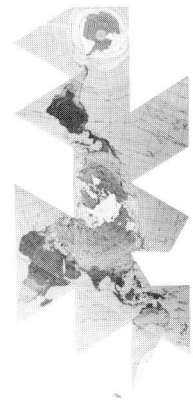

Regular icosahedral Dymaxion sky-ocean map. By Fuller.

Japanese commemorative stamp showing a Fuller dome serving as a weather bureau on the summit of Mt. Fuji.

Chinese dog with a geodesic sphere, before Fuller. Summer palace, Beijing. Eighteenth or nineteenth century.

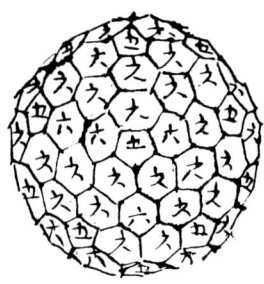

Geodesic sphere drawn by Yasuaki Aida, before Fuller. From Sanpo-Kiriko-Shu (A Mathematical Collection of Polyhedra). Eighteenth or nineteenth century.

and an egg hatches a chicken, not a butterfly. All this takes place through synergy.

Fuller's ancestors were traders, plying the ocean for many generations, thus Fuller himself was familiar with the ocean. To show that the surface of the terrestrial sphere is loosely occupied by the ocean, he invented a polyhedral terrestrial globe, in opposition to traditional cartography, and wrote *Operating Manual for Spaceship Earth*. His globe usually has the shape of the regular icosahedron or cuboctahedron, because it is more easily developed onto a plane than is the sphere. As a result he was aware of spherical patterns having a powerful impact and being beautiful in appearance.

Always advocating peace, happiness, and prosperity of all mankind, and asserting that the world of the future would be more inhabitable than it is today, Fuller started to design huge regular icosahedral or dodecahedral vessels for citizens. These are the geodesic domes or Fuller Domes or Bucky's Bubbles, which remind us of Plato's regular dodecahedral vessel of the cosmos. Their design is almost fixed, as shown in figure ⑮.

One is derived from subdivision of faces of a spherical regular icosahedron by geodesic lines which are parallel to the edges, as at the right side of the figure. The other is obtained by subdivision by perpendiculars to the yellow edges, which are eliminated when all the perpendiculars are drawn, as in the central sphere. The largest sphere in the figure is decorated with the foregoing two patterns.

In practice, each type of triangle becomes smaller and more numerous, and the radius becomes larger. In the near future, it will become a large complete sphere and will fly high in the air. Already a small one has been put at the summit of Mt. Fuji, the highest place in Japan, as if it were threatening to fly upward.

When the Spaceship Earth has become too wasted to live on, one night people will have to build some geodesic spheres having the diameter of about 1 kilometer, thinking that the last day of the earth draws near. The work, if people unite their strength, will be easy, according to synergy, because it consists of combining some small and light units. After finishing the work, people will get into their spheres and will await the next morning. Just at the moment of dawn, sunlight will come into the spheres to warm the air. Then the buoyancy of the air will cause the spheres to rise and fly high, carrying people as in castles in the air. Figure ⑪ shows the gallant sight. To make a passing reference, the largest and dark sphere is a photo of its inside. The life in the spheres will be much better than the present and the people will be intoxicated with peace, happiness, and prosperity.

Polyhedra, which were a Pandora's box for Kepler, are a Noah's Ark for Fuller.

The Mandala

6

by Kukai

The regular polyhedra, which according to Kepler have done much for harmony in the world, must have greatly influenced the Orient, which forms one part of that harmonious world.

Ancient Chinese around the time of Christ wrote in *Chun-Nan-Tzu, Chuang-Tzu,* and elsewhere, that the word "Yu-Chou" ("universe" in English) consists of "Yu" meaning 3-space and "Chou" meaning time.

Concerning the form of the universe, there were two typical opinions: the heaven is circular and the earth square, both spreading in parallel planes; and they are concentric spheres. The circle and the sphere are the most fundamental figures, even if this was not taught by Pythagoras.

Ancient Chinese said that the Chinese continent had been created by two gods with snakelike bodies, Fu-I and Nu-Kua, probably a man and his wife. According to many reliefs portraying them, Fu-I always has a compass and Nu-Kua a carpenter's square as tools for designing the heaven and the earth.

Fu-I holding a compass (right) and Nü-Kua holding a carpenter's square (left).

The Yin-Yang theory of that time says that the compass, which is a circle, suggests the male, heaven, sun, and the like, and the carpenter's square, which is a square, the female, earth, moon, and so on. And the five Chinese elements of the universe—wood, fire, earth, metal and water—were born to them.

The ancient Japanese, who learned much from the Chinese, seemed to be influenced by such a myth. For instance, the word "Ten-En-Chi-Ho" meaning the circular (En) heaven (Ten) and square (Ho) Earth (Chi) was frequently used to explain the universe. The word "Ki-Ku," meaning the circle (Ki) and square (Ku), was also usually used for architectural techniques. An ancient burial mound, "Ko-Fun," built for an emperor or empress from around the fourth to the seventh century (the Kofun era in Japan), has a precise geometrical shape derived from the circle and the square as in figure 12.

"Goshikizuka" (five-colored mount) Kofun, Kobe. Reconstructed in the twentieth century.

A gigantic one having a square and circle in a row is the most typical. The largest example is the tomb, built for the Emperor Nintoku, with an area of about 500,000 square meters. As for their curious shape, there is a suggestion that the circle and square are arranged from the right to the left to represent heaven, the world after death, and the Earth, the world before death.

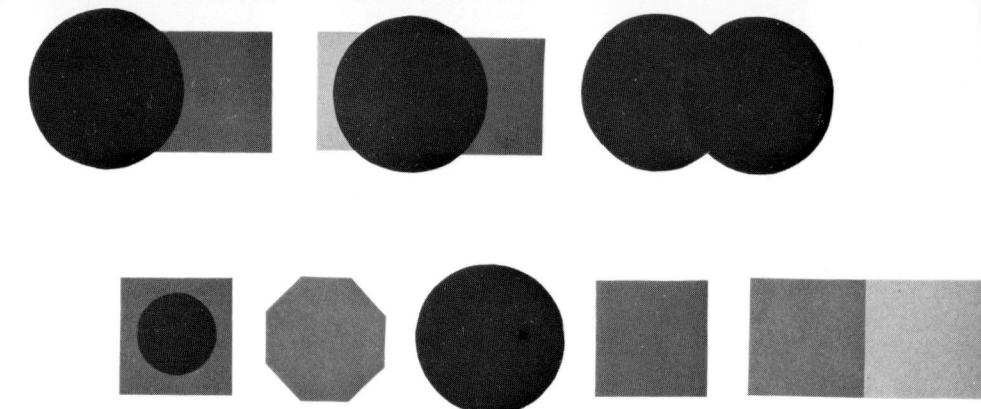

Plans of Kofuns, ancient burial mounds in Japan.

Laputa-island in Gulliver's Travels.

Kaaba at Mecca.

During the Kofun era Buddhism had not yet reached Japan, but the religious life of the people had already begun. Later under the influence of Buddhism, their religion emerged as Shintoism. Believers in Shintoism had shrines as the place of prayer, and the god was often symbolized by a circular or regular polygonal, usually regular octagonal, mirror, or sometimes by a spherical object, such as a temari. Even now, in the main halls of shrines, such objects representing the god are placed at the center. Further, a circular mirror and a straight sword are always set at the right and left as if to suggest Fu-I and Nu-Kua.

Buddhism, which originated in India at the same time as Pythagoras, was introduced into Japan via China in the middle of the sixth century. A famous ruler of Japan at that time, Prince Shotoku, who some say was born in a manger like Christ, encouraged Buddhism as a governmental enterprise. He carried a carpenter's square in some portraits drawn by later artists. Almost all of the temples commemorating him have had the regular octagonal plan seen in the picture of the 10,000-yen note of today's Japan. (Unfortunately, it was decided to change the design in 1985.)Through Shotoku's influence, Buddhism prospered and many priests traveled to China to study it at first hand.

Kukai (Kobo-Daishi) in the ninth century, a priest, artist, novelist, scientist, and more, was the most famous among them. He went to China to study Esoteric Buddhism, which had originated in India in the second century, and upon his return propagated Shingon Esoteric Buddhism (the Shingon sect). The word "Shingon" means that all truth lies in languages or forms of things.

The Shingon sect became prevalent among the Japanese under the protection of the emperors of the time and many temples were built as places of worship. Among them, two-storied pagodas, called "Tahoto," which have the square plan for the lower floor (the Earth) and the circular plan for the upper floor (heaven), are even now typical landmarks for every Buddhist pilgrim, and not merely for those of the Shingon sect.

The cosmology of Esoteric Buddhism originally appeared in the Kosá-Sāstra scripture compiled in India in the fifth century. It runs as follows. There is a vast circular expanse of ocean on the top surface of a giant cylinder as in figure ⑯, whose diameter reaches about ten times that of today's sun. In the center of the ocean, there stands a cubic land, or Oriental paradise, the Sumeru, which is thought to symbolize Mt. Everest. It is surrounded by seven quadrilateral mountain ranges whose width and height become greater as they draw toward the Sumeru. Around the mountain ranges there are four islands: a semicircular one in the east, a regular triangular one in the south, a circular one in the west, and a square one in the north. Our human race lives on the regular triangular island, located in India to the south of Mt. Everest, which has seven cubic levels of hell hanging one below the next under the ground. In the sky, just above the Sumeru, seven square islands for the gods, all of whose shapes are congruent with a face of the Sumeru, float like the floors of a seven-storied building, exactly like the Laputa-island inhabited by geometry-minded people in *Gulliver's Travels*. The sun, moon, and stars revolve around the floating islands like kites describing circles in the sky. When men on the triangular island see the sun disappear behind the Sumeru, night falls.

Altar of Heaven. Beijing. Nineteenth century.

Such a realm is merely a primal world, or solar system. The whole universe is composed of 10^9 of such worlds. Buddha is outside the universe like the Trinity described by Dante.

Interestingly enough, the Sumeru and the four islands on the sea are shaped like the five (not four) elements of the universe inherited from ancient India: the Sumeru as the cube (like the Kaaba at Mecca, which is the most sacred place of worship for the Moslems) and the western circular, southern regular triangular, eastern semicircular, and northern square islands symbolize earth, water, fire, air, and heaven respectively. According to *Hsi-Yu-Chi*, Sun-Wu-Kung, a Chinese monkey, symbolized by the red fire, runs as fast as light around these five islands.

Buddha modeled after a Gorinto pagoda.

The urban plan of the present capital of China, Beijing, represents the very image of such a geometrical universe, though Chinese people originally believed in a different set of five elements, as previously mentioned; there is the Palace Museum in the center, whose plan is nearly a square, and around it are the Altar of Heaven, having the circular plan in the south; the Altar of the Earth, having the square plan in the north; the Altar of the Sun, having the circular plan in the east; and the Altar of the Moon, having the semicircular plan in the west.

The ancient capitals of Japan, at Nara and Kyoto, might be planned like Beijing, though one cannot be sure, for they are buried underground. Nevertheless, a small five-storied pagoda, a "Gorinto," which has been standing on earth since the beginning of the twelfth century, when the capital was Kyoto, clearly shows the universe in relation to the Indian five elements. Gorinto are made of stone, wood, copper, crystal, and so forth, and even now "live" as tombstones, vessels

Tomb stone of Chogen as a trigonal Gorinto (right), thirteenth century, and a usual Gorinto (left). Todaiji temple, Nara.

Womb Mandala (upper) and *Diamond Mandala* (lower). *Drawn by Kaiun Tatebe, a modern priest. Twentieth century.*

of holy things, or as memorial monuments for dead persons. A Gorinto consists of five blocks, whose shapes are cubic, spherical, square-pyramidal, hemispherical, and chestnutlike from bottom to top. Generally, Sanskrit letters or Chinese characters are inscribed or carved on each block, naming earth, water, fire, air, and heaven, always in the same order. Sometimes, the blocks are colored yellow, white, red, black, and blue, always in that same order.

There are various opinions as to why such a composition was adopted. One is that it was modeled after a pagoda of ancient India, and the other is that it was modeled after the human body, from the bottom to top: a sitting leg, abdomen, chest, face, and head.

Some historians, including Mayumi Takizawa, express another view, that the Gorinto represent a stacking of the five Platonic solids as shown in the central pagoda of figure ⑰, from bottom to top: a cube, regular icosahedron, regular tetrahedron, regular octahedron, and regular dodecahedron.

Chogen, an eccentric priest; whose birth in the twelfth century coincided with the appearance of an early gorinto, designed Todaiji Temple. Nara, the largest wooden building in the world. He admired a so-called trigonal Gorinto, which has an exact regular tetrahedron in the place of the block representing fire. His own tombstone at Todaiji Temple is also a trigonal Gorinto.

A shadow picture of such a Gorinto, made from a thin board, is called a "Stupa" and is used even now as a simplified gorinto. Some examples are shown at both sides of figure ⑰. They usually have no color, unlike those in the picture, and their shapes are longer than those in the picture.

If a Stupa is taken apart to yield five geometrical shapes, these may be arranged like the Sumeru and the four islands, in figure ⑯.

The Mandala, introduced into Japan by Kukai as an important altar piece of Esoteric Buddhism, also includes such plane figures. The word "Mandala" means a collection of the essences of things or elements of being. It had its origin in India as a round plot of ground, but it was brought into China and Japan as cosmic tableaux in which the Great Bear and other constellations were sometimes painted.

The most formal Mandalas are worshiped in pairs. One is the "Mandala of the Womb World," representing the actual and physical world proper to the female, and the other is the "Mandala of the Diamond World," representing the ideal and mental world proper to the male. These are usually hung on either side of the image of the patron saint in such a way that the former is on the right or east and the latter on the left or west.

The Womb Mandala looks like a plan of the site of a big temple or city such as Beijing. In the middle of the central section, inscribing an eight-petal lotus, sits Vairocana, suggesting the sun, and revered as the nucleus of the universe.

It is said that this section represents the element earth, a yellow cube. At the left the goddesses of mercy hold lotus flowers. They are sometimes said to relate to water, a white regular icosahedron. Above Vairocana a small regular triangle, as the heart of Buddha, is always painted to stimulate the worshiper's intellectual curiosity. It may represent the element fire, a red regular tetrahedron. At the right of Vairocana, fierce-faced multiarmed Buddhas scold the worshipers so they will become spiritually awakened as soon as possible. They may be designed to play the part of the element wind, a black regular octahedron.

Finally, below Vairocana, the universe or heaven, which holds the other four elements, is presented. The Buddha sitting in the center sometimes has a close packing of four chestnut-shaped gems. His name, Akasāgarbha, means a warehouse containing nothing, like the vessel of the universe, a blue regular dodecahedron. The two Buddhas placed on both sides of him have either 100 or 1,000 hands in which various things composing the universe are seen. Their hands may be too few to construct the whole of the universe.

On the other hand, the Diamond Mandala is composed of nine sections, all of which are dominated only by squares and circles. Each section imitates the central one at whose very center Vairocana sits surrounded by four Buddhas. This arrangement is repeated persistently from the part to the whole, with various meanings and teachings. It is difficult to understand it all perfectly. For this reason, in the three sections of the uppermost row, a few typical Buddhas are drawn in order that laymen may understand more easily. There is no Buddha except Vairocana in the center of the row.

Almost all of such Buddhas are, curiously enough, related to the five elements (the Platonic solids). If so, the "Regular Polyhedra Mandala" would be derived as in figure ⑱, where each Buddha is replaced by corresponding polyhedron. Some empty circles indicate the places of Buddhas who cannot be associated with five elements no matter how hard we try.

The Mandalas are usually hung in the darkness of the sanctum, and are seldom exhibited before the public. A "Kiriko-Doro," a cuboctahedral lantern, secretly illuminates such Mandalas. A "Kiriko" generally meant a polyhedron in early Japan, but it came to indicate only a cuboctahedron in later ages. According to a custom handed down from about the thirteenth century, this lantern is hung in the most important place in temples, and only at the Bon festival, which is held annually in mid-August to commemorate the dead. No one knows the reason, but in Korea, the nearest country to Japan, a rhombicuboctahedral lantern is quite important. According to Heron, Plato studied the cuboctahedron along with the regular polyhedra. If so, some serious attention might also well be given to the former. Overwhelming the power of Vairocana, who shines more brilliantly than the sun, the influence of Plato extended both to heaven and to hell in Japan.

Thousand-armed Buddha. Toshodaiji temple, Nara. Eighth century.

Kiriko-Doro-lanterns used at the Bon-festival of Akimoto-shrine, Kyoto. 1982.

Rhombicuboctahedral lanterns used at a Bon-festival in Korea. 1982.

7 Synergy

by Fuller

Sketch of snowflakes by Toshitura Doi, an ancient lord of Japan. From Sekka-Zusetsu (Pictorial Collection of Snowflakes). Nineteenth century.

Japanese traditional crests showing regular pentagonal flowers. Plum blossoms (upper four) and cherry blossoms (bottom four).

Man, fascinated by the infinitely large universe, is equally intrigued by the infinitely small world of molecules, atoms, and elementary particles, and tries to grasp all in a totality.

Priests of ancient India, who imagined the macro universe in the Kosa-Sāstra scripture in the fifth century, also had interests in the micro world. It is said that the bottom left section of the Diamond Mandala shows that the magical power of Vairocana extends even to the microscopic world. According to the scripture, all matter in this world is made of minute corpuscles of indistinct figures, neither circles nor squares, and invisibly small.

First, seven such corpuscles constitute a mutually perpendicularly crossed cluster like three coordinate axes in 3-space as shown in figure ⑲. Next, seven such clusters are united in a similar manner, constituting a larger cluster having 49 corpuscles. If we repeat the same construction seven times in all, then about 820,000 corpuscles form a large tree shape. At this stage, it becomes visible as one of the elements of the universe.

Such a microscopic world can also be derived from spheres, just as Pythagoras derived his microscopic world from circles.

For instance, philosophers of the Vaisésika school, which flourished before Christ, adapted the shapes of the elements to spheres and not to regular polyhedra, and in variations of the arrangement of spheres they found various aspects in nature.

Kepler's poor eyes, which looked toward the far end of the macrocosm to find his cosmic cup, were one day drawn to snowflakes, falling from that end of the macrocosm. His treatise on snowflakes, written at the beginning of the seventeenth century, is said to have initiated experimental observations about atoms and molecules of matter.

The treatise, entitled *On the Six-Cornered Snowflake*, was a New Year's gift to his patron, who admired "nothing" such as snowflakes, which, in melting, vanish so quickly into "nothing." In his article, Kepler tried to clear up the riddle of why snowflakes have regular hexagonal shapes, while most flowers, though equally beautiful, are regular pentagonal. His conclusion was that he did not know the reason, and that he would rather listen to the reader's opinions.

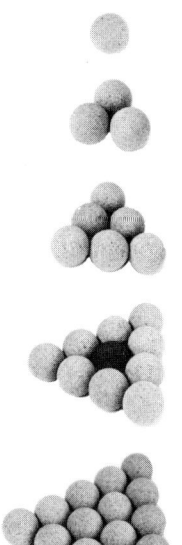

[13] Basic construction of the universe by Fuller.

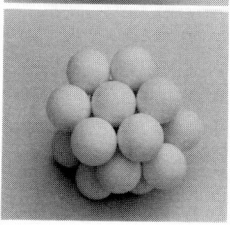

Two kinds of the closest packing of spheres. The position of the third layer differs in each.

Honeycomb using only one kind of 18-hedra (lower) and a development of its unit. According to Heinz-Dieter Löckenhoff and Erwin Hellner.

Honeycombs of regular and semiregular polyhedra. Similar units have constant color. Each in the left half is derived as follows: those of the top row from regular polyhedra only, those of the second and third rows from regular polyhedra and Archimedean solids, those of the fourth and fifth rows from Archimedean solids only, and those of the bottom row from Archimedean solids and Archimedean prisms. Those marked * are Andreni's honeycombs. Each in the right half is derived from an Archimedean prism, including the cube (top to fourth row) or from four kinds of units including pseudo-rhombicuboctahedra.

Structure of A (left column) and B quanta module (right column) of Fuller. The top row shows the development of each and the second row the outer appearance with its mirror image.

Kepler assumed two types of planar arrangement of equiradial spheres formed by drops of aqueous vapor. In one, neighboring spheres formed regular triangles. In the other, they formed squares. He stacked these planar arrangements in two layers and obtained two kinds of sphere arrangements in which each sphere contacts at least four other spheres: the cubic-lattice arrangement and the closest packing.

Among all arrangements of spheres, the latter has the largest number of spheres in a unit volume, as the name indicates. Figure [13] shows the composition in the form of the Pythagorean tetractys. He then says that the sphere arrangement which relates to snowflakes must be derived from the cubic-lattice type. This is because drops of warm vapor should float freely up in the sky, maintaining highly symmetric relationships with each other and having voids of as similar a shape as possible, due to God's warm heart. The closest packing has irregular shapes of voids.

Subsequently, the cool air of the upper sky invades the voids and eats away at each drop, starting on the side nearest the largest void, until only three mutually perpendicular diameters of each drop remain, as directions for drifting 3-space. And as soon as they fall onto the ground the drops will lie along the three diagonals of the regular hexagon. Judging from these surmises, the earth seems to prefer a cube, Plato's earth, or hexagon.

Among flowers, a lily, with its bulb deep in the earth, blooms in a regular triangle or hexagon. And well above ground, an apple blossom or flower of the cucumber blooms in a regular pentagon. Kepler acknowledges that, among

㉑ ㉒ ㉓

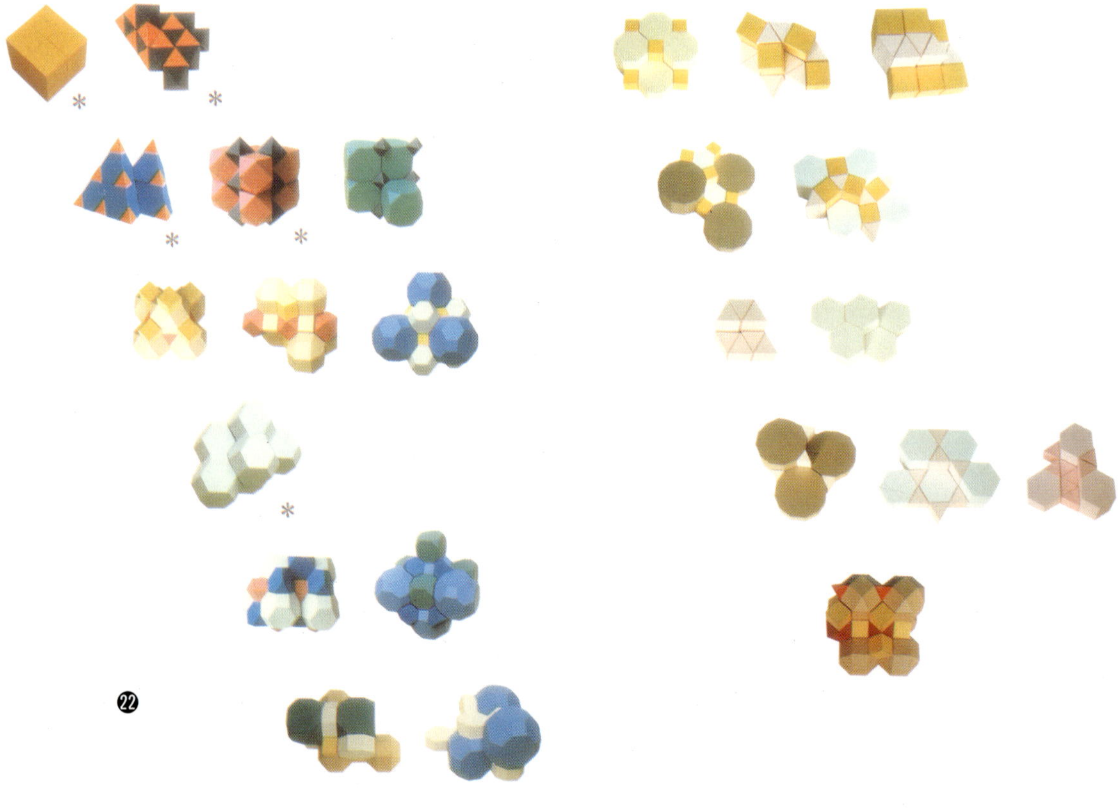

42

㉔

㉕

㉖

[14]

Rhombic dodecahedron (upper) *and rhombic dodecahedral lattice.*

The loosest packing of spheres. A regular tetrahedral sphere arrangement is put at each vertex of a rhombic dodecahedral lattice.

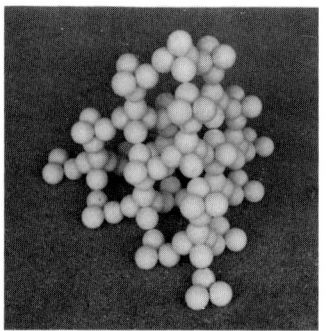

Designs of art and nature in Fuller's universe.

◀ ㉕

Construction of space-filling polyhedra from quanta modules. Top and second rows show various space fillers, the third row a truncated octahedron (right) and polyhedra appearing in it; the bottom row, from left to right, a polycube, its section, an octet-honeycomb, and its inner structure.

Plato (left) *and Aristotle, created en route to space-filling by quanta modules.*

crystals of minerals which lie hidden deep in the bowels of the earth, there are also some having regular pentagonal faces, such as pyrite. It is clear that we must make more many-sided, or "polyhedral," observations.

Though real honeycombs or seeds of pomegranate have regular hexagonal sections, they seem to relate to the closest packing and not to the cube type. This is because, at the closest packing, tangent planes of each sphere at points of contact constitute a stacking of rhombic dodecahedra as shown in figure [14], each of whose units has 12 congruent rhombi with diagonal ratio $1 : \sqrt{2}$. It becomes seeds of pomegranate by itself, or a real honeycomb as shown in figure [15], when it is elongated along its six sets of mutually parallel edges.

A cube and its dual, a regular octahedron, can be inscribed in this rhombic dodecahedron as in the center of figure [16], so a rhombic dodecahedron may also relate to snowflakes. On the other hand, a regular dodecahedron and its dual, a regular icosahedron, can be inscribed in a rhombic triacontahedron as in the right of figure [16]. This polyhedron has 30 congruent rhombi with diagonal ratio $1 : 1.618$. It may relate to regular pentagonal flowers. The remaining part of figure [16] shows a cube in which there is inscribed Kepler's stella octangula, a figure which embraces self-dual regular tetrahedra. It may relate to stars.

15
Model of a natural honeycomb.

27 ▶
Universe of polyhedra.

28 ▶
Stackings of saddle polyhedra. The group colored blue on the left is derived from a cubic lattice, one colored green in the center from an octet-truss, and one colored red, at the right, from a rhombic dodecahedral lattice. The upper row shows the polyhedral (right) and facial (left) units. After Michael Burt.

29 ▶
Two examples of toroids having a tunnel, which are derived from convex polyhedra having regular polygonal faces, as seen in 12. Outer and inner appearances of each are shown to the left and right. After Bonnie Stewart.

Summing up, the Cube family seems to be related to hexagonal snowflakes, and the Dodecahedron family to pentagonal flowers. Nevertheless, snowflakes actually spread in a plane, rather than as faces of spatial polyhedra. Why? This is probably because regular hexagons cannot be put together to form convex polyhedra, no matter how numerous they may be. A snowflake is thus obliged to spread in a planar shape of the Pythagorean regular tessellation of hexagons. But if so, why is the size of snowflakes not the same?

All may be God's works.

Kepler left a clue to the experimental observation of the microscopic world of molecules or atoms, and this idea has been growing little by little like a soap bubble. At last, his view of nature based on closest packing became the origin of the ultramodern ideas of Fuller, which are even now inspiring vast geodesic domes, or "Bucky's Bubbles."

According to Fuller's *Synergetics*, the whole universe can be understood as a closest packing of minute equiradial spheres of masses of energy. In particular, one regular tetrahedral portion derived from five close-packed layers as in figure 13, is thought to be a fundamental form in nature. Like the tetractys of Pythagoras, it has a heart in the center. If further spheres in the closest packing are added on, polyhedral layers of 12, 42, 92 spheres, and so on, surround the central one. The outer appearance of each layer always becomes a cuboctahedron as in figure 17.

The universe is expanding at a tremendous speed, always having the shape of a cuboctahedron, a "vector equilibrium" whose distance from the body center to each vertex is equal to each edge length. The outer shape of the universe for Fuller is not a regular dodecahedron but a cuboctahedron. We refer

Molecular structure of DNA devised by Fuller. This perverse DNA may be of the Fuller family.

㉗

㉘

㉙

46

㉛

㉚

㉛

16

Mutually dual regular polyhedra inscribed in a cube, rhombic dodecahedron, and rhombic triacontahedron, from left to right.

the skeptic to the microcosmos of the heaviest modern element in nature, the structure of uranium. It should have three layers of the closest packed spheres around the central one as in figure 17 so that 92 spheres form a cuboctahedron in the outermost layer.

Therefore, the atomic number of uranium is 92, and the number of the neutrons, which is the total number of spheres, not counting the central one, is 146 (= 12 + 42 + 92), and the atomic weight is 238 (= 92 + 146).

More complex features in nature are explained as follows: the closest packed 13 spheres, when they are arranged in a cuboctahedral shape, have the heart in its center. If the heart is removed, the end of living nature draws nearer. First, the remaining 12 spheres are placed at 12 vertices of a regular icosahedron. May not dangerous viruses be born in this way? Second, each of six pairs of them is reduced to one sphere, twelve becoming six, and they are put at six vertices of a regular octahedron. May not inorganic diamonds be crystallized in this way? Further, each of four triplets of the previous icosahedral arrangement is reduced to one sphere, twelve becoming four, and they are put at four vertices of a regular tetrahedron to form the essential element of nature. Ergo, according to ideas of synergetics, nature can always be explained by integers, which are numbers of spheres. Nowhere can irrational numbers be seen. From the beginning, nature must have been created as simply as possible. The molecular formula of water is H_2O, not $H_\pi O$. The irrational or transcendental number, such as $\sqrt{2}$ or π, is only a monster born from an invented system, mathematics. A point should be considered as an infinitesimal made up of regular tetrahedron having four spheres, so a straight line is actually a zigzag line made up of such regular tetrahedra stacked in a row. In the ensemble, we have something that resembles our idea of a light ray,

 ㉚

Final stage of the stellation of a regular dodecahedron, presented in a "cross-eyes" stereogram.

 ㉛

*Stellations of regular dodecahedra. Those marked * are the original regular dodecahedra, ● small stellated dodecahedra, ▼ great dodecahedra, and ■ great stellated dodecahedra. Scattered small pieces appear on the way to the stellation.*

with its characteristic wave length. Two of such lines intersect each other at a single regular tetrahedron as a point. Therefore, 1 + 1 = 4.

The coordinate axes appropriate to Fuller's universe form an octet-truss which is derived from the closest packing, using the above mentioned lines to connect the centers of the spheres. Sixty degree angles and regular triangles, therefore, play a leading part in his universe. The system of orthogonal coordinate axes, dominated by 90 degree angles and squares, is merely a scaffold for a castle in the air, mathematics.

Figure ⑱ shows an octet-truss of one layer which covers the earth and the thinnest possible "cloud" of spheres, which covers the octet-truss. The thinnest "cloud" means two layers of the closest packing of spheres. It has the least thickness capable of shutting out parallel sunbeams of constant direction. Even a cloud is ruled by the closest packing and 60 degree angles.

If this is so, how should the voids among the closest packed spheres be treated in the "Fuller" universe? Fuller, never puzzled, says that in the voids, there are concave regular octahedral and cuboctahedral bodies, some of whose faces are pieces of surfaces of equiradial spheres turned inside out as in figure ⑳. Spherical cuboctahedral chains appear on each surface of the closest packed spheres, fastening them so they do not fall apart.

The shape of the universe is truly a cuboctahedron, not a regular dodecahedron.

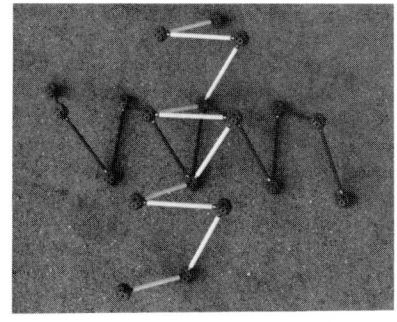

1 + 1 = 4. According to Fuller, a line has wave linearity like a sun ray, and if two lines intersect mutually, they form a regular tetrahedron.

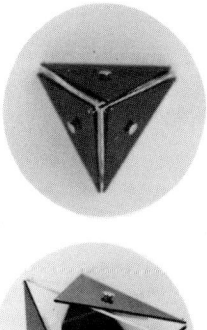

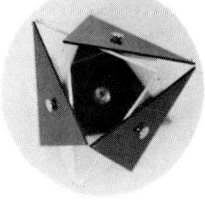

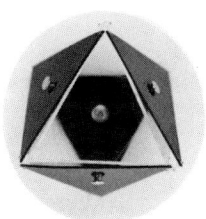

Transformation of a regular tetrahedron to a regular octahedron as Fuller's jitterbug. By Joe Crinton, a pupil of Fuller.

[17] Construction of a molecular structure of uranium, after Fuller. From left to right, the central sphere, 12 of the first layer, 42 of the second layer, and 92 of the third layer. They always have a cuboctahedral appearance.

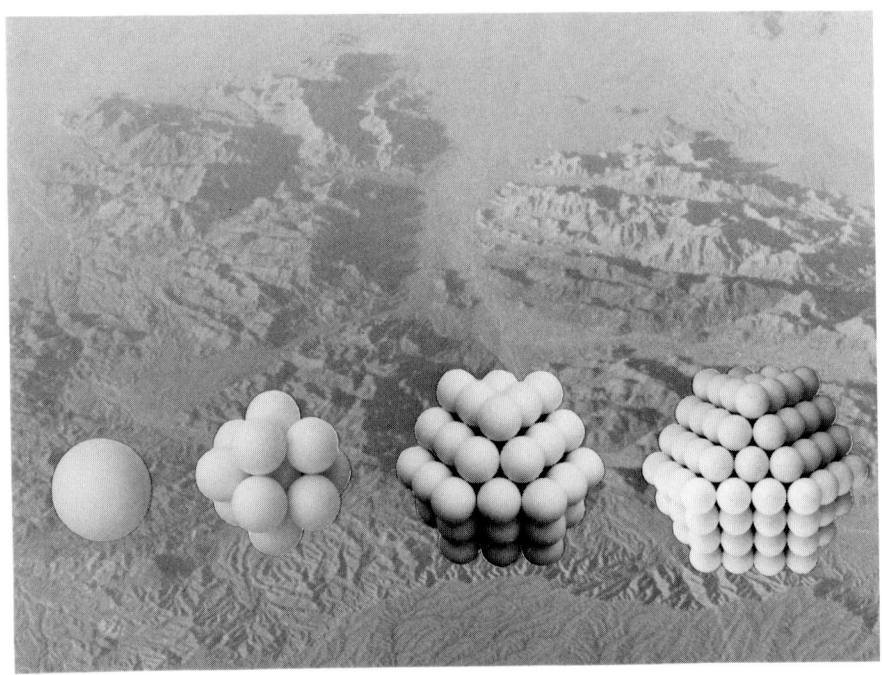

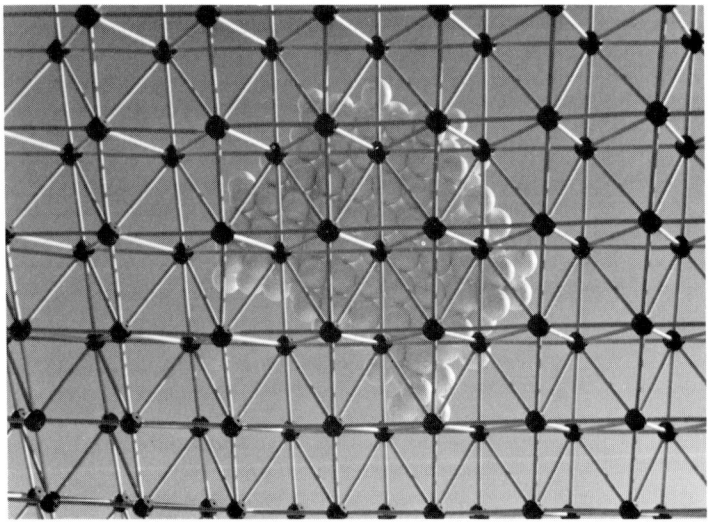

[18] "Thinnest cloud" covering an octet-truss.

The Palace by Kelvin

8

Polyhedra are more stable in a group than they are separately. Some scientists, especially mineralogists and chemists, have used space-filling polyhedra as a better means to analyze nature. Figure ㉑ shows a space-filling using only one kind of 18-hedron. It looks like a stone wall, as seen in the countryside. Such an infinite stacking of polyhedra, packed face-to-face and filling space without gaps, is called a honeycomb, and polyhedra which can construct a honeycomb by themselves are called space-fillers.

It seems that Pythagoras was already aware of two kinds of honeycombs which use only the Platonic solids: a polycube using only cubes, and an octet-honeycomb using regular tetrahedra and regular octahedra.

If Archimedes were to construct a universe without voids, using polyhedra, what shapes would be possible? Figure ㉒ shows various honeycombs derived from regular and semi-regular polyhedra. Each type of unit is consistently colored. The five marked by asterisks are sometimes called Andreni's honeycombs in which the placement of polyhedra is the same around every edge. Of the space-fillings in figure ㉒, that using only truncated octahedra, seen in the fourth row on the left, has the least surface area, given a constant volume. It can be directly derived from the closest packing of spheres, if each sphere is inflated without moving its center. Such features were studied by Lord Kelvin; the truncated octahedron is occasionally called Kelvin's Solid. According to the most recent *Guinness Book*, Kelvin entered a university at the youngest age ever (when 10 years old). In other words, he enclosed the largest amount of knowledge within the least cranial surface.

The space-fillings in figure ㉒, however, are too numerous for a simple universe. E. S. Fedorov, a modern crystallographer who mastered elementary algebra at the age of five, condensed this list of space-fillers to only five parallelohedra, using the fact that all forms in figure ㉒ have a period of translational symmetry.

Convex space-fillers, which can fill 3-space using only parallel translation and placed face-to-face, are called parallelohedra.

There are five and only five parallelohedra, shown in figure ⑲: a cube, regular hexagonal prism, truncated octahedron

Jungle gyms as honeycombs regular tetrahedra and truncated tetrahedra (upper), or regular octahedra and cuboctahedra (lower).

  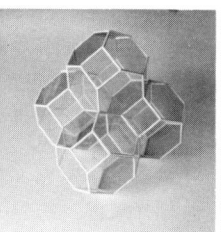  

19

Parallelohedra. From left to right, a cube, regular hexagonal prism, truncated octahedron, rhombic dodecahedron, and elongated rhombic dodecahedron.

(Kelvin's solid), rhombic dodecahedron, and elongated rhombic dodecahedron. All of figure ㉒ can be classified into these five types according to their translational symmetries.

The shapes of parallelohedra are variable. For example, a cube can be transformed into any parallelopiped; a regular hexagonal prism into any oblique hexagonal prism whose bases are parallelohexagons. However, no matter how they are deformed, the following features are invariant: their faces are always parallelogons, polygons in which opposite sides are parallel and of equal length. Each pair of opposite faces are congruent and lie on parallel planes. All parallelohedra are symmetrical about their body centers. Each set of mutually parallel edges forms a zone surrounding the polyhedron, like a set of crossties on a railway.

More generally, polyhedra having such properties are called zonohedra, whether they are space-fillers or not. Zonohedra can easily and reasonably be adapted to designs in nature, because of their familiar features.

Regular polyhedra, with the exception of the cube, are not zonohedra, so perhaps there is something unreasonable about Plato's view of nature. To improve on this situation and grasp nature more rationally, Fuller devised two fundamental space-fillers by subdividing a rhombohedron, one of the zonohedra.

The stoicheia of Fuller's universe are two kinds of tetrahedra, called A and B quanta modules. The A module, shown in the left column in figure ㉓, is created when a regular tetrahedron itself part of the fundamental rhombohedron for octet honeycomb, is cut into 24 equal parts along planes of symmetry. It has a mirror image as in the second figure from the top, and a plane development shown at the top. The triangle placed at the center of this development coincides with Plato's scalene stoicheia.

The B module, shown in the right column in figure ㉓ is created when a regular octahedron, itself also a part of the fundamental rhombohedron for the octet honeycomb, is cut as follows: first, construct a flat tetrahedron using six A modules, as in the second figure from the bottom of the left column. Next, remove such tetrahedra from the faces of a regular octahedron, so that a solid eight is constructed, seen at the second from the bottom in the right column. Finally, cut the

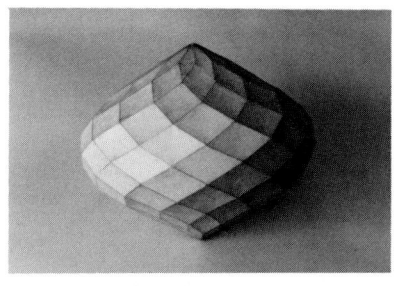

Zonohedron having 132 faces.

remaining solid into 48 equal parts along its planes of symmetry. The resulting module has a mirror image as shown in the second figure from the top, and a plane development shown at the top. The triangle placed at the center of the development coincides with Plato's isosceles stoicheia.

Thus quanta modules resemble stoicheia. But, stoicheia are only plane figures and can construct only two-dimensional features of the universe. Quanta modules, containing three-dimensional inner space, assure a "fuller" future in contrast with the empty past as constructed of stoicheia. Beautiful forms in nature, such as the pentagonal flowers and hexagonal snowflakes in figure ㉔, can be built from quanta modules. They do not, however, have space-filling shapes.

Various space fillers can be obtained as in figure ㉕ when suitable quanta modules are combined face-to-face. They are "Mite," "Bite," "Rite," "Lite," "Kate," "Kat," and "Octet" from left to right in the top row, and "pseudo-octahedron," a cube, and three kinds of rhombic dodecahedra from left to right in the second row. The right end of the third row shows a truncated octahedron. The others in this row, from left to right, are Kepler's stella octangula, a regular compound of a cube and a regular octahedron, and a cuboctahedron, which are not space-fillers but appear in the process of building the truncated octahedron.

The bottom row shows a polycube and an octet-honeycomb with their sections. The figures in figure ㉖, which are obtained in the process of stacking of quanta modules, resemble the portraits of Plato and Aristotle in the *School of Athens* painted by Raphael. There is a rumor that his Plato was modeled on Leonardo da Vinci and his Aristotle on Michelangelo.

In any case, quanta modules, or synergy, hold the key to the birth of great genius.

School of Athens. The Vatican. Sixteenth century. Painted by Raphael. Plato is at left of the center with his Timaeus *pointing to the heaven, and Aristotle at right of Plato with* Etica, *pointing to the ground. Around them, many geniuses of ancient Greece, such as Pythagoras, Socrates, Archimedes, and also Raphael himself.*

9 The Equation
by Euler

Annular toruses of the Erechtheion, Athens

According to Kepler, the universe appears to have the simplicity of polyhedra. Polyhedra, however, have complex aspects, just like the universe. The previously mentioned regular and semiregular polyhedra are merely a few of the convex polyhedra, just as the planets of the solar system are only a few of those in the galaxy. Similarly, the convex polyhedra are merely a few of the more general polyhedra, just as our galaxy is only one of many in the universe. Such a rambling world of polyhedra is briefly classified in figure ㉗.

The whole universe of polyhedra is shown in a large yellow ellipse. In it, there are two star clusters colored in yellow ochre. One, indicated in a circle, represents those infinite polyhedra which are supposed to have an infinite number of faces, and the other, indicated in an ellipse, represents the finite polyhedra having a finite number of faces.

The finite polyhedra have two star systems colored in brown. One, indicated in a circle, is of the open polyhedra which have at least one edge belonging to only one face, and the other, indicated in an ellipse, of the closed polyhedra all of whose edges belong to two and only two faces.

The closed polyhedra have two planets colored in vermilion. One, indicated in a circle, is of the compound polyhedra which have arbitrarily many tunnels, and the other, indicated in an ellipse, is of the simple polyhedra which have no tunnels. If the number of tunnels of a compound polyhedron is infinite, it will coincide with an infinite polyhedron. On the contrary, if the number is only one, it is called a torus-shaped polyhedron due to its similarity in appearance to a torus, such as those carved on the capitals or bases of columns of ancient temples.

The simple polyhedra have two satellites colored in carmine. One, indicated in the right circle, is of the concave polyhedra which have at least one dihedral angle of more than 180 degrees, and the other, indicated in the left circle, is of

Euler's equation for convex polyhedra. V means the number of vertices, E of edges, and F of faces.

$$V - E + F = 2$$

[20]

Regular sponges. From top to bottom, around each vertex, having six square faces and six windows, four regular hexagonal faces and two square windows, and six regular hexagonal faces and six regular triangular windows.

[21]

Sponge-shaped infinite polyhedra. From top to bottom, derived by gathering around each vertex four regular octahedra, a regular icosahedron and two regular octahedra, a regular icosahedron and a regular octahedron.

the convex polyhedra all of whose dihedral angles are less than 180 degrees.

Among such various polyhedra, the convex polyhedra, which were mentioned last, are so widespread that they are merely called polyhedra. There are, however, infinitely many kinds of convex polyhedra, and it is almost impossible to grasp all of them. For example, there are 2, 7, 34, and 257 kinds of convex pentahedra, hexahedra, heptahedra, and octahedra, respectively. It may be impossible to calculate the numbers of convex polyhedra with more than eight faces.

Surrounded by such vast and obscure satellites of convex polyhedra, Plato constructed a universe using only regular polyhedra, which is symmetrical and holy, as if it were just born directly from the hand of the Creator.

Coxeter's three regular sponges, shown in figure [20], have similar qualities of symmetry and beauty. In regular sponges, regular polygons are fitted together in the same number around each vertex, leaving congruent regular polygonal gaps, that is, windows. They extend infinitely in 3-space beyond the known universe and separate all the space into two equal parts. There are many other sponge-shaped infinite polyhedra, made using only regular polygons, but the shapes of their windows are no longer regular. Figure [21] shows some examples, all of whose outer shapes can be derived from face-to-face stackings of regular polyhedra.

|22|

Infinite polyhedron separating 3-space into two equal parts. After Michael Burt. It consists of squares, rhombi whose two diagonals have the ratio 1 : $\sqrt{2}$, and regular hexagons.

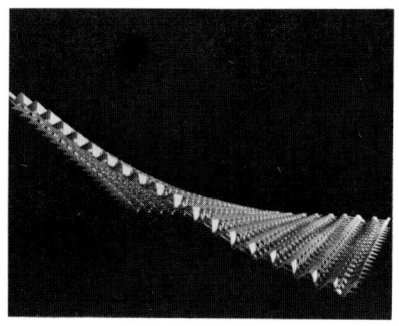

Two kinds of hyperbolic surfaces. Hyperbolic paraboloid made of a metal deck devised by Kuniaki Ito (upper) and a hyperboloid of revolution of one sheet as an observation tower, Kobe (lower).

Pythagorean regular tessellations, Platonic regular polyhedra, and Coxeter's regular sponges represent the typical features of three well-known geometries of possible worlds, each characterized by the sum of face angles around a vertex. In the case of regular tessellations, the sum of the angles around each vertex is equal to 360 degrees as required in the planar Euclidean space. In the case of regular polyhedra, it is less than 360 degrees, and *we are* in spherical Riemannian space. In the case of regular sponges, the angle sum is more than 360 degrees, and we find ourselves in hyperbolic Lobachevski-Bolyai space.

Newton said that the universe extends infinitely, as does Euclidean space. On the contrary, Einstein emphasized that it is closed, as is Riemannian space. Perhaps some day in the future, a genius may insist that the universe is a Lobachevski-Bolyai space. Gauss, who seriously studied the Riemannian and Lobachevski-Bolyai spaces, forecast that the real 3-space may coincide with that of Lobachevski-Bolyai. He surveyed the sum of internal angles of one large triangle whose vertices were at the summits of three mountains, thinking that if the sum were to be less than 180 degrees, his conjecture would be confirmed. He failed in the surveying because a triangle on the Earth, however large it might be, is still too small to provide information about the shape of the whole universe.

Notwithstanding, if Gauss's surveying had met his expectation, the Earth would have been thrown into utter confusion, as if the ground and the sky had been reversed. Universes of regular sponges spread infinitely, and the Earth and the sky,

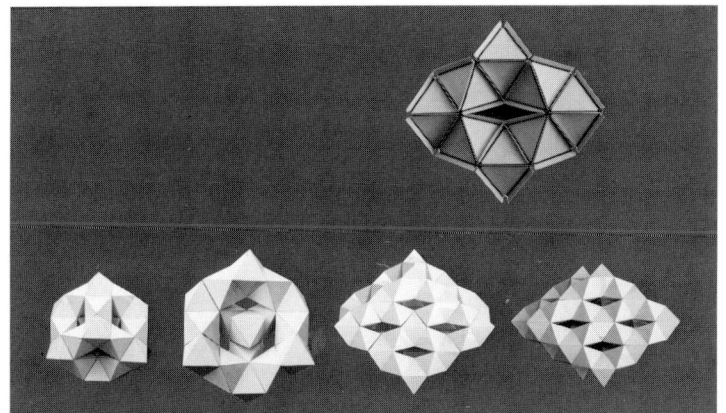

Toroids derived from regular octahedra. After Bonnie Stewart. A unit (upper) and its various stackings.

Stackings of regular compounds whose units are shown at the top. Made from origamis. The left is constructed by using Kepler's stella octangula and the right by using compounds of a cube and regular octahedron.

One-sided polyhedra derived from cuboctahedra. A 10-hedron (left) and 12-hedron.

One-sided heptahedron (center) and stackings of one-sided polyhedra. At the left is the case using heptahedra and 10-hedra, and, at the right, that using heptahedra and 12-hedra. The faces of any one color construct infinite planes.

which are congruent to one another, repeatedly exchange their positions.

A monkey, Sun Wu-Kung, in *Hsi-Yu-Chi*, found a lively world of goblins who lived in stone caves, and in this world there were other stone caves, each containing another inhabited world. There are many similar tales in the *Arabian Nights*, *Alice in Wonderland*, and in other books of that sort. Heroes or heroines of such fictional tales might also live on an earth in the shape of a sponge, that is, in Lobachevski-Bolyai space.

Figure 22 shows more exotic sponge-shaped infinite polyhedron by M. Burt, a modern Israeli architect who lives surrounded by arabesques. This sponge separates 3-space into two equal parts, as do the regular sponges. It has squares, regular hexagons, and rhombi of diagonal ratio $1 : \sqrt{2}$.

P. Pearce called such infinite polyhedra "labyrinths," even if they do not divide 3-space into two equal parts. Labyrinths are derived from various lattices as follows. First, in a space lattice or relative to a combination of such lattices, stretch saddle-shaped minimal surfaces (saddle surfaces) between lattice lines which are skew to one another, so as to obtain stackings of one kind of saddle polyhedra, whose faces are saddle surfaces. Then divide their faces into minute approximate polygons.

The bottom row in figure 28 shows three stackings of saddle polyhedra derived from a cubic lattice (left), an octet-truss (center), and a rhombic dodecahedral lattice (right). On the right of each upper row is the saddle polyhedral which is derived from facial units shown at the left.

Such a curious infinite sponge-shaped world, like a labyrinth, will only be found in a future which is infinitely far away. In the near future, however, a world having finite tunnels may be discovered. J. A. Wheeler, a modern physicist,

Jungle gym in the form of Pearce's labyrinth. Chokoku-no-mori Museum, Hakone, Shizuoka.

57

㉜

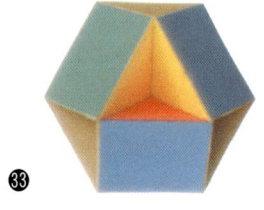

㉝

㉞

㉟

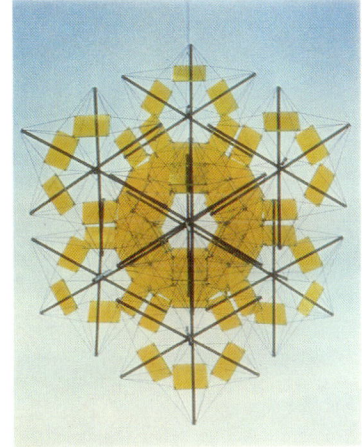

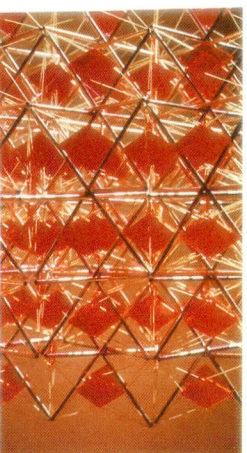

㊱

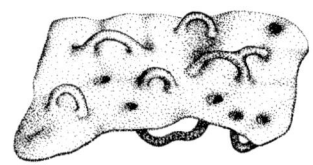

Scene of a super-micro world. According to John Wheeler.

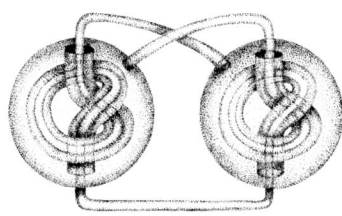

Topological surface having genus two, which has the simplest shape next to spectacles. It rather looks like an anatomical chart of eyes, not of spectacles.

says that in a microscopic universe, whose sun is an atomic nucleus, time and space are warped geometrically and unidentified space having "worm holes" and "bridges" will be found as a compound polyhedron having many tunnels.

B. Stewart, a modern geometer, makes polyhedra with regular polygonal tunnels, calling them toroids, and describes adventures in their world. For example, he combined assemblages of equal height of convex polyhedra having regular faces shown in figure ⑫, and made various toroids as in figure ㉙. The inner and outer shapes are shown at left and right, respectively. The number of tunnels is arbitrary.

Further, he devised a torus-shaped polyhedron having 48 regular triangles as the outer surface of a stacking of eight regular octahedra, as at the top of figure 23. This ring can be stacked so that the outer shape is a compound polyhedron having many tunnels. There are two types of spatial extension shown at the bottom right in figure 23. If regular dodecahedra or regular icosahedra are used instead of regular octahedra, the same construction can be carried out.

It is often said that animals who have digestive organs, including man, can be topologically replaced with compound polyhedra having some tunnels. Plato, who constructed nature out of regular polyhedra, was not aware of this, though he often looked at the toruses of the Erechtheion. If he had known of the relation between compound polyhedra and the human body, he would not, by any means, have revised the *Timaeus*. This is because there is no "compound regular polyhedron" whose faces are congruent regular polygons, the same number fitted together around each vertex.

Computer graphics of a combination of 12 regular hexagonal prisms. By Mamoru Hozaka and Shizuo Shimada, who use an original word "GEOMAP."

Regular polygonal panels in tensegrity structures. By Harriet Brisson.

The Labyrinth

by Möbius

10

Heavenly bodies, which were sometimes given the shapes of regular polyhedra, are seen to glitter as stars in the distance.

When faces of a convex polyhedron are extended little by little, they expand until they intersect each other and form a family of successively larger star-shaped polyhedra. Such an operation is called the stellation of a polyhedron, and the newly born star-shaped polyhedra are called stellated polyhedra.

The faces of a regular dodecahedron, the outer shape of the vessel holding Plato's cosmos, can be extended until they finally form the figure ㉚. Figure ㉛ shows an intermediate form. Those marked with *s show the original regular dodecahedra, those marked with ●s are the small stellated dodecahedra (the first stellation), those marked with ▲s are the great dodecahedra (the second stellation), and those marked with ■s are the great stellated dodecahedra (the final stellation). The small pieces scattered in the figure are obtained along the way.

According to Coxeter et al., 58 stellated polyhedra can be derived from a regular icosahedron. The left-hand figure of the top row in figure ㉔ shows one such polyhedron obtained at the final stage. Many geometers pay special attention only to the great icosahedron shown at the right in the same figure. That is because only four, namely this one and the three stellated dodecahedra just mentioned, are analogues of regular polyhedra, as follows. The small stellated dodecahedron has five regular pentagrams around each vertex. The great dodecahedron, the great stellated dodecahedron, and the great icosahedron have respectively five regular pentagons, three regular pentagrams, and five regular triangles around each vertex. These four, therefore, are called the stellated regular polyhedra.

In contrast to regular polyhedra, which had been known before Christ, stellated regular polyhedra are relative newcomers to the world. The star-shaped polyhedra drawn by Leonardo da Vinci, shown in Figure ⑩, do not belong to the above class of stellated polyhedra. They are star-shaped merely because regular polygonal pyramids adhere to their faces.

Star-shaped polyhedra in Daniele Barbaro's La Pratica della Perspettiva, *1569. The top right one resembles Kepler's great stellated dodecahedron.*

Metamorphoses of three kinds of Yoshimoto cubes.

Metamorphoses of three kinds of polyhedra rotating like smoke-rings. All of them are made of only one kind of tetrahedron having yellow, red, blue, and green right triangles.

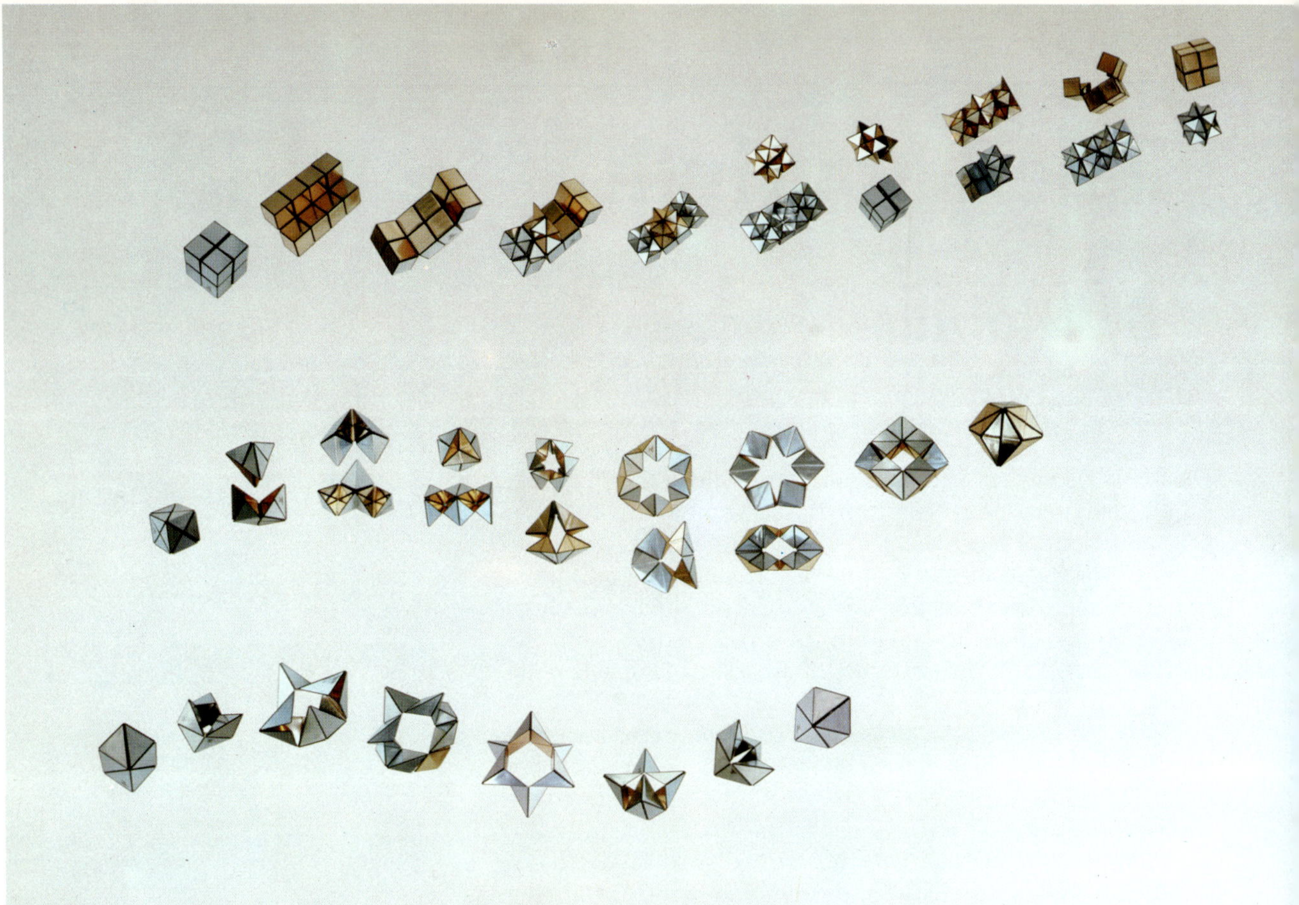

37

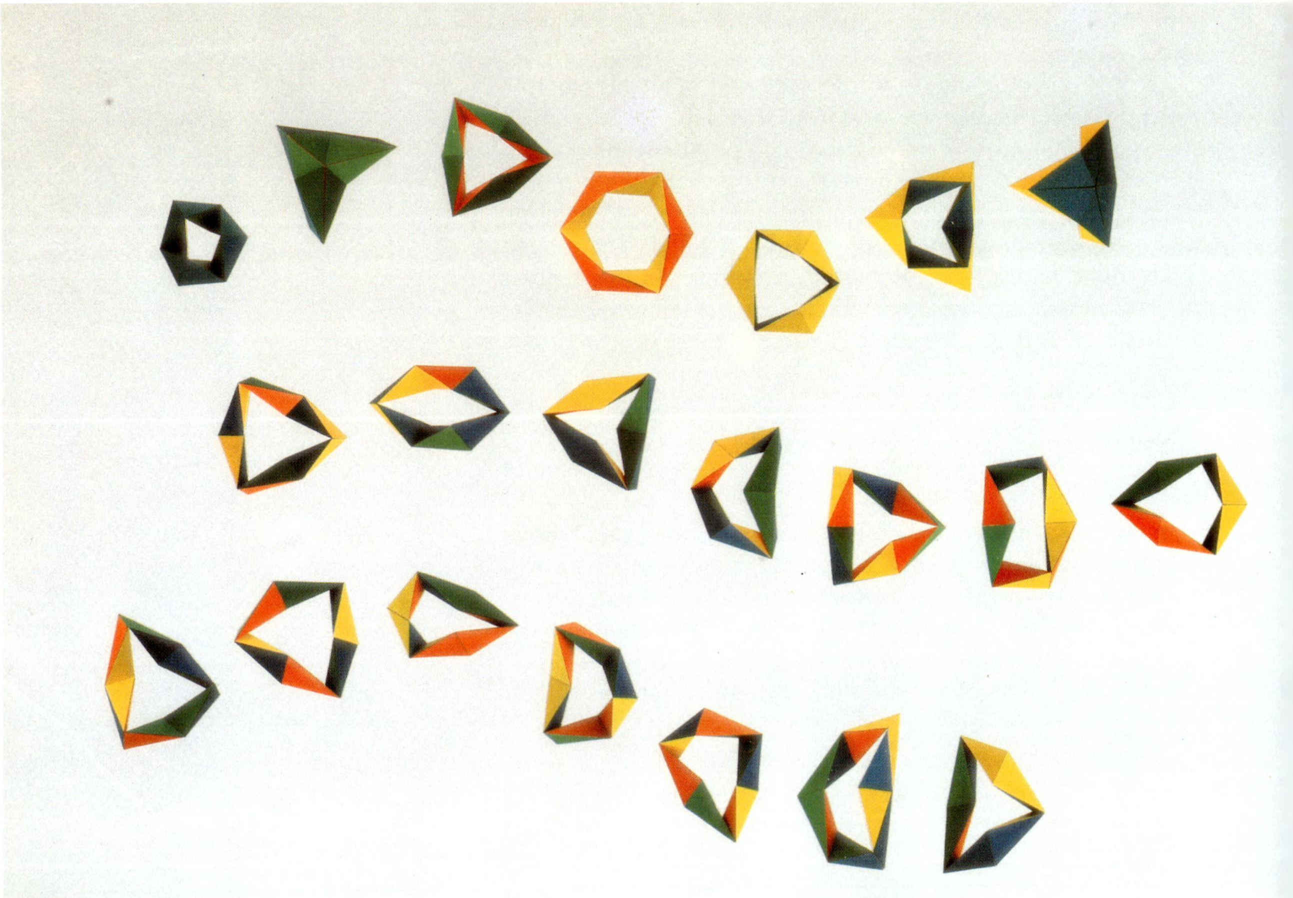

38

㊴

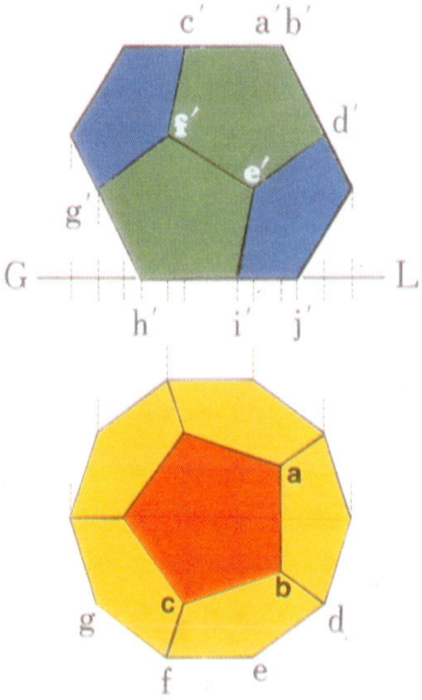

㊵

㊶

63

⟨24⟩ ▶

Stellated polyhedra. The final stellation of a regular icosahedron (top left), *the great icosahedron* (top right), *regular compounds* (two rows in center), *a chiral regular compound* (bottom left), *and a combination of two regular dodecahedra.*

City hall of Bat-Yam, Israel. Designed by Zvi Hecker.

◀

Modern arabesque derived from regular polygons. After Roger Penrose.

◀

Plan (below) *and elevation* (upper) *of a regular dodecahedron.*

◀

Planar arrangements of four kinds of pentagons seen in 40.

Kepler, probably for the first time in human history, drew pictures of the small stellated dodecahedron and the great stellated dodecahedron in *Harmonices Mundi*. He said that the former, a hedgehog, might be inserted in his cosmic cup, combining a regular dodecahedron and icosahedron with the Earth between. About 200 years later, L. Poinsot found the great dodecahedron and great icosahedron. The four stellated regular polyhedra are sometimes called the Kepler-Poinsot solids.

Now, what kinds of stellated polyhedra are born from the remaining regular polyhedra?

None of these are born from the regular tetrahedron or from the cube, but Kepler's stella octangula, in which three regular triangles fit together around each vertex, is born from the regular octahedron. This is a compound of two regular tetrahedra. It is called an orthogonal regular compound in which two mutually dual regular or stellated regular polyhedra are combined so that pairs of corresponding edges intersect at right angles.

There are five regular compounds, including Kepler's stella octangula, as in the center of figure ⟨24⟩. One, which is born between a small stellated dodecahedron and a great dodecahedron, coincides with the great dodecahedron.

There are also five chiral regular compounds in which some congruent regular polyhedra are combined so that all their vertices can coincide with those of a regular or semiregular polyhedron. These are as follows: Kepler's stella octangula, five tetrahedra in a regular dodecahedron, as shown in the left of the bottom row of figure ⟨24⟩, ten tetrahedra in a regular dodecahedron, five cubes in a regular dodecahedron shown roughly in figure ⟨30⟩, and five octahedra in an icosidodecahedron. The right-hand figure in the bottom row of figure ⟨24⟩ was obtained by combining two regular dodecahedra; it has no special meaning.

Among these polyhedra are some examples which can be stacked as if they were a star cluster. For example, the left and right halves of figure ⟨32⟩ show stackings of Kepler's stella octangula and orthogonal regular compounds derived from cubes and regular octahedra respectively, all of which are made as origami. Two regular compounds in the upper row are the units of the stackings. The former shows an octet-truss and the latter a combination of an octet-truss and a polycube, both of which are useful for architectural designs.

Figure ⟨25⟩ shows a planar design using the stellation of a rhombic triacontahedron by Z. Hecker, a modern Israeli polyhedric architect. Such a design is used in architecture by Hecker and covers the earth as a star, as if there is the distant future in the distant past.

Stellation refers to the outward extension of polyhedra. On the contrary, the inward "intension" of polyhedra produces one-sided polyhedra. Polyhedra having faces whose front and

Planar representation of the stellation of a rhombic triacontahedron. By Zvi Hecker.

Möbius strip.

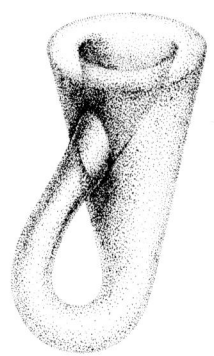

Klein bottle.

rear side cannot be distinguished, like the famous Möbius strip, are called one-sided polyhedra, though they are not included in figure ㉗ because of their unusual nature.

Some of them can be derived by inward modification of regular or semiregular polyhedra. For instance, in a cuboctahedron, if its eight regular triangles are removed, and instead of them, four regular hexagonal "diagonal planes" are added as new faces, then a one-sided 10-hedron is obtained, as shown at left in figure ㉝. In this case, the intersecting lines of the diagonal plane should not be thought of as edges. The front and rear side of each face can no longer be distinguished, because they are now contained in a Möbius strip.

Similarly, a one-sided 12-hedron can be derived from a cuboctahedron by removing six square faces and adding four diagonal planes as shown at the right in figure ㉝. One further famous one-sided polyhedron is a so-called one-sided heptahedron which is derived from a regular octahedron by removing four regular triangular faces, each pair having only a vertex in common, and replacing them with three square diagonal planes as shown in the center of figure ㉞. The left and right sides of figure ㉞ show stackings of the above mentioned one-sided 10-hedra and heptahedra on one hand, one-sided 12-hedra and heptahedra on the other hand, both of which correspond to honeycombs of regular octahedra and cuboctahedra. As a result, if each face has a suitable color, 3-space is divided by numerous parallel multicolored planes, as if they were floors, walls, and ceilings of a building. It seems that unusual one-sided polyhedra are useful even for conventional architectural designs.

A. Holden, a modern chemist, devised other star-shaped solids having no face. These are called "polylinks," and they

Polylink. By Alan Holden.

use straight rods which are joined at their ends into regular polygonal rings. Figure ㉖ shows an example in which ten regular triangular rings are linked through one another in the form of an icosidodecahedron, which makes it rigid.

These complicated stellated structures have also been designed by computer. Figure ㉟ shows an example by M. Hozaka, a modern space scientist. However, the brilliant starlike constructions shown in figure ㊱ could not be designed by computer. They were designed by H. Brisson, a modern polyhedric artist. She first constructed an arbitrary lattice using straight rods, then stretched some fine threads tightly between certain pairs of rods, to form a spider's web and to support some regular polygonal panels, as if they were food for the spider.

Fuller gave to the principle in such design the name "tensegrity." It means a structure in which some delicate tension members (women) act with integrity and exhibit enormous power to support thick compression members (men). This is in accordance with Fuller's idea of synergy.

The tensegrity structure must not move, though it looks ready to move and may vibrate. N. Yoshimoto, a modern polyhedric sculptor, however, created truly movable sculptures. Figure ㊲ shows three types of the Yoshimoto cube. The top row shows the metamorphosis in which a silver cube having a golden inner unit changes to two cubes, one silver, one gold, each lacking an inner unit. Along the way, stellated polyhedra born from rhombic dodecahedra appear.

The middle row shows a cube formed from two rings, each of 12 tetrahedra, derived from the equipartition of a cube into 24 pieces. The bottom row shows a cube formed from one ring of 12 tetrahedra derived from equipartition of a cube into

Tensegrity mast. Smithsonian Museum, Washington.

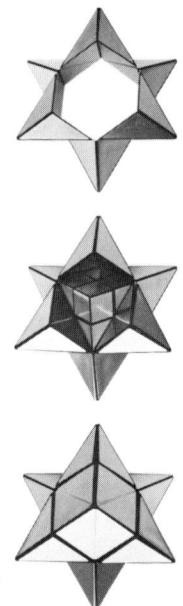

Combination of three types of Yoshimoto cubes. The uppermost is seen in the third row of thirty-seven, the middle is a combination of that and one of the top row of thirty-seven, and the bottom is a combination of all three.

12 pieces. These three types of Yoshimoto cubes can be curiously combined into a single star-shaped body.

The figures in figure ㊳ which resemble the latter two Yoshimoto cubes show other movable ring-shaped polyhedra. They, three kinds in all, can be rotated from the position on the left to that on the right, and back to that on the left, the way a smoke ring rolls. Each ring is similarly constructed of six congruent elongated tetrahedra joined edge to edge. These faces are only two kinds of right triangles.

In summary, honeycombs, whose component polyhedra are connected with each other face-to-face, are stable, while the Yoshimoto cubes or rotating smokerings, whose component polyhedra are connected with each other by their edges, are movable, and individual polyhedra are free. As mentioned by E. S. Fielitz, a modern polyhedric architect, they may represent the three-phase system in nature: solids, liquids, and gases. Even if they are strange polyhedra, they also relate to nature.

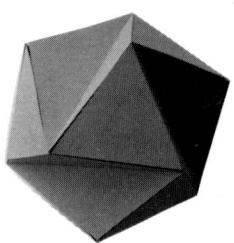

Movable polyhedron derived from a regular icosahedron.

Magic by Dürer

11

Plato's stoicheia and Fuller's quanta modules, which can construct nearly everything in the universe, dislike regular pentagons and pentagrams. The regular pentagon and pentagram, however, have been favored by some people because of their beauty and the presence of the golden ratio 1 : 1.618.

It is said that the ancient Egyptians and Greeks used this ratio in the Great Pyramid, the Parthenon (designed by Phidias after whose name the number 1.618 is oftenly called "ϕ"), and others. As already described in detail by Euclid, the golden ratio appears in a regular pentagon and pentagram as AC : AB, AG : AC, AH : AB, AJ : AH, in figure 27.

Pythagoras associated good health with a pentagram, and used it as the logo for his school. He intended to keep secret the existence of a regular dodecahedron, all of whose faces are regular pentagons. Hippasus, notwithstanding, disclosed the secret and was punished by the gods, drowning in the Mediterranean. Despite this incident, one finds today, off the coast of Japan, women divers looking for pearls wearing hats embroidered with amulets in the shape of pentagrams.

There are about 50 nations in the world whose national flags are decorated with pentagrams: The Stars and Stripes of the United States, the Soviet flag, the Five-Golden-Stars-in-the-Red-Sky of China, and many others. The Star of David of Israel, a hexagram, and the complete circle of the Rising-Sun flag of Japan, are rather the exceptions.

Not only on Earth but also on the moon, where there is no life, regular pentagons and pentagrams are scattered. When interplanetary probes land softly on a planet, the Soviet Union usually puts regular pentagonal insignia on it. Already, a regular dodecahedral insignia was delivered and the United States has hoisted its Stars and Stripes.

The territory of regular pentagons differs essentially from that of regular hexagons, as the United States differs from the Soviet Union. Kepler thought that a pentagon occurs in organic matter and a hexagon in inorganic matter. He skillfully classified the Archimedean solids into the Cube family relating to regular hexagons and the Dodecahedron family relating to regular pentagons.

National flags having only pentagrams. From top to bottom, Morocco, Somalia, the Republic of China.

Hexagram, a Star of David, seen in the national flag of Israel.

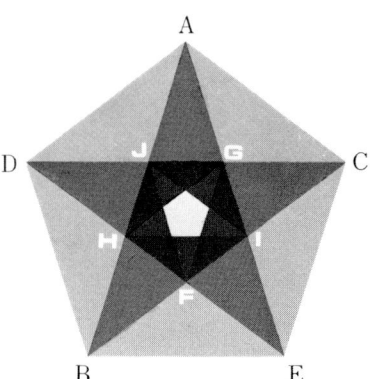

[27]

Regular pentagons and pentagrams.

Melancolia I. *Etching by Albrecht Dürer. 1514.*

Model of the truncated rhombohedron in Melancolia I. *Reconstructed by Kazuko Emoto.*

All members of the Cube family are ordinary and sociable. Almost all of the already mentioned tessellations and honeycombs belong to this family. However, in the course of time, it is the Dodecahedron family, possessed of extraordinary creative powers, which has come to be paid the keenest interest.

In arabesques, regular pentagons or pentagrams were arranged though they had to be delicately warped as in figure ①. In the Renaissance, many-sided geniuses, such as L. Pacioli, who wrote *De Divina Proportione*, intensely studied the golden ratio in its relation to a regular pentagon. Dürer, who left us *Melancolia,* was one of them. An angel Melancolia, represented as a young woman architect with a compass, is said to symbolize an Architect of the Universe, while various things around her suggest melancholic temperament. The large sphere may be melancholy because it has the outer shape of the mysterious universe. But why is a large truncated rhombohedron melancholy? It makes one melancholy not to be able to understand it.

Recently, Kazuko Emoto, a Japanese woman sculptor, reconstructed this pictorial polyhedron in its original state, using the golden ratio. She says that each of the pentagonal faces is obtained when the rhombus ADBI in figure [27] is cut with two lines passing through F and G, both parallel to DI. As if in response to her expectation, Dürer himself left various patterns of regular pentagons in *Unterweisung der Messung,* as shown in figure [28]. Of them, the second from right in the bottom row is probably the first development of a regular dodecahedron in history. A more interesting figure is shown at the right end in the same row. It is a defective circle made of nine regular pentagons; no more pentagonal units can be inserted into the gap.

Kepler, however, later devised a pattern as shown in figure [29] and published it in *Harmonices Mundi.* There are many circuits of 10 regular pentagons.

R. Penrose has favorably regarded such regular pentagons and, though he is a future-minded mathematician, he devised a curious odd-looking arabesque as in figure ㊴.

Patterns of regular pentagons only. After Dürer.

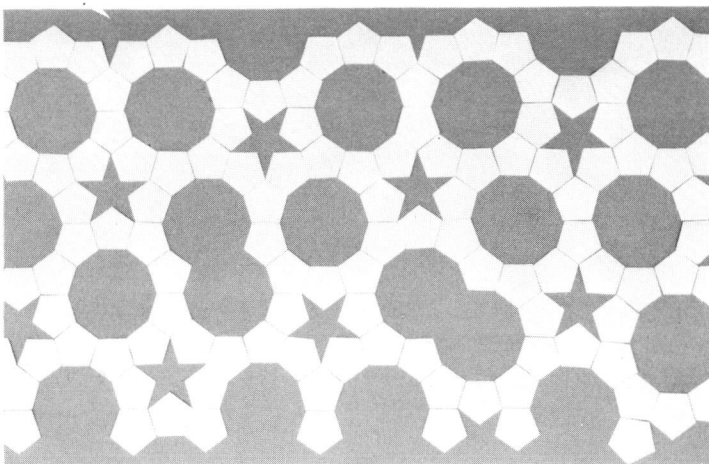

Pattern of regular pentagons only. After Kepler.

Such eccentricity of regular pentagons suggests something abnormal going on in the Dodecahedron family. A regular dodecahedron as the father can be stacked by parallel translation with edges or parts of faces in common, as in figure 30. Stacked this way, they seem to submit to the traditions of the Cube family. Z. Hecker designed an apartment house in a desert in Israel after the model of such stackings. He thinks that, because comfort and calm are necessary for a residence, the comfortable manners of the Cube family must be adopted.

On the contrary, if one wishes to enjoy leisure, the eccentric stackings as in figure 31 are rather suitable. The upper stacking, where adjacent polyhedra have edges in common, resembles Kepler's tessellation in figure 29. The lower one, figure 31, shows how 20 small regular dodecahedra are inscribed in a larger regular dodecahedron. If each unit, one small dodecahedron, is replaced with a newly born large regular dodecahedral stacking, a much larger regular dodecahedral

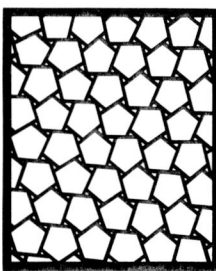

Chinese lattice derived from regular pentagons.

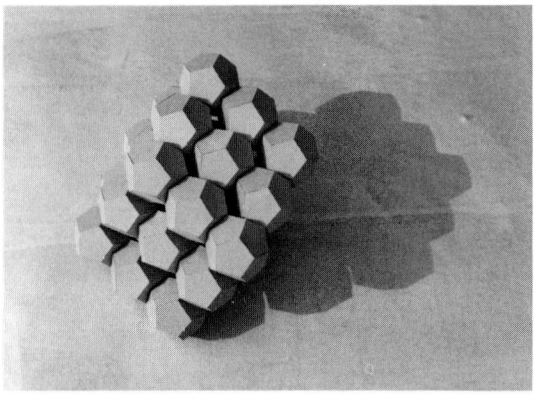

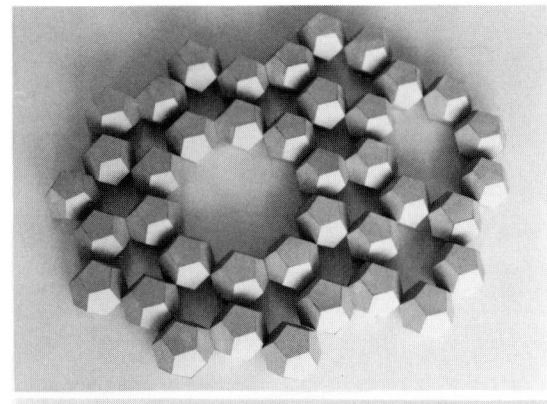

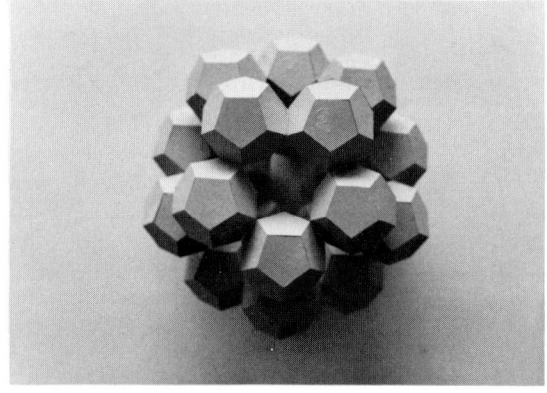

[30]

Periodic stackings of regular dodecahedra.

[31]

Nonperiodic stackings of regular dodecahedra. The upper one covers Kepler's pattern in [29], and the one below is a regular dodecahedral stacking of 20 small regular dodecahedra.

Ramot housing. Jerusalem. Designed by Zvi Hecker.

stacking will be created. As a result, Plato's regular dodecahedral universe is infinitely expandable. Curiously enough, in either stacking in figure [31], there is no period of translation, as is usually the case in the Cube family.

What is the orthogonal projection of such mysterious regular dodecahedra? Figure ㊵ shows a plan (below GL) and an elevation (above GL) of the regular dodecahedron. In the plan, the red regular pentagonal face abc is placed in such a way that one of the edges, ab, is perpendicular to GL. Next the point d is obtained as the intersection of the line bd, which passes through the center of the red regular pentagon, and the line cd, which is parallel to GL. The points d, e, f, and g become the vertices of a regular decagon.

In the elevation, the lines e'g', d'f', and a'c' are parallel to GL. The distance between e'g' and GL, which is equal to the distance between d'f' and a'c', is equal to the radius of the circumcircle of the red regular pentagon in the plan, and the distance between e'g' and d'f' is equal to bd, the radius of a regular decagon in the plan. The points a', h', and j' become the vertices of a parallelohexagon.

According to such a troublesome construction, some curious tessellations and stackings of pentagons and dodecahedra can be derived as follows. As shown in figure ㊶, all four kinds of pentagons, shown in red, yellow, green, and blue, have the same largest width, and they can be arranged in a plane, though various gaps remain and the periods of parallel translations are seen. Examples using only red regular pentagons were already mentioned.

Furthermore, in 3-space, these pentagons can produce five pentagonal dodecahedra, including a red regular dodecahedron as in figure ㊷. Of them, the upper four can be clustered without inner gaps, with faces of the same color in common as in figure ㊸. A red regular dodecahedron is in the center. The whole outer shape becomes a convex 42-hedron of 12 red regular pentagons and 30 parallelohexagons having the same appearance as the elevation in figure ㊵.

While regular dodecahedra are considered as fathers, regular icosahedra may be thought of as mothers. A mother is a woman who is beautifully dressed up and rouges her lips. Figure ㊹ shows two stackings of regular icosahedra, meeting edge to edge and dressed up with lip-shaped units, as shown in the upper figure. Like her husband, she can also be arranged with periods of parallel translations visible as at the left. But the figure on the right is unusual. It looks like the regular dodecahedral stacking shown at the bottom of figure ㉛, because 12 small regular icosahedra make up a large regular icosahedron.

In each regular icosahedron, we can inscribe three mutually perpendicular golden rectangles, whose edge ratio is the golden ratio, such that their 12 vertices coincide with the vertices of the original regular icosahedron. Using this method the unusual constructions in figure ㊹ can be transformed to the usual constructions in figure ㊺.

The pentagon with the golden ratio, which brought mysterious order out of chaos in the past, will give a clue to reasonable order in the chaotic future.

Three mutually perpendicular golden rectangles inscribed in a regular icosahedron.

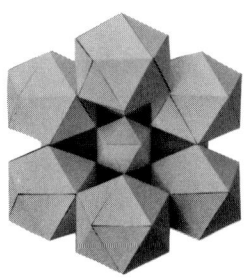

Small regular icosahedron embedded in a regular icosahedral cluster of 12 large regular icosahedra seen in the right of forty-four. The ratio of edge lengths of the small and large icosahedra is the golden ratio.

Five pentagonal dodecahedra derived from four kinds of pentagons seen in 40.

Finite face-to-face stacking of four kinds of pentagonal dodecahedra seen in the top row of 40. A red regular dodecahedron is in the innermost place of the one on the left.

Periodic rhombohedral (left) and nonperiodic icosahedral (right) stackings of regular icosahedra. Each model is made of lip-shaped units shown at the top.

Representation of each of 44 by golden rectangles.

12 The Planet

by Penrose

Nature has an inclination to disorder itself so as to increase entropy. Nonperiodic patterns of the pentagon and the Dodecahedron family illustrate this law.

However, there are always some gaps between the units, and they are not suitable for the nongap universe. Are there nonperiodic patterns having no gaps? In the 1930s Voderberg devised a nonperiodic tessellation in which seahorse shapes, concave nonagons, are tessellated infinitely in a double spiral without any gap, as in figure ㊻. Under close inspection, we see the sea horses are arranged so irregularly that we may well wonder whether the double spiral can grow endlessly or not. The possibility has, however, been established.

By way of contrast, Golomb's sphinx, which was discovered a little later, can doubtlessly be arranged nonperiodically without any gap, because it is always stacked in a sphinxlike shape, with elementary pieces of the same shape, as in figure ㊼.

Everyone will be able to find such nonperiodic tessellations if not forced to use only one kind of unit. Figure ㊽ shows an example which was found by a Japanese schoolboy. Geometers, dissatisfied because examples of nonperiodic tessellations are so numerous that even schoolboys can find them, chose to add a severe condition: no matter how skillfully a

Voderberg's double spiral.

Golomb's sphinx.

Japanese boy's car.

Essentially nonperiodic tesselations by Penrose. The upper one is built of dart shapes and kite shapes and the lower one of two kinds of rhombi on which white lines and parts of little white and red circles are marked.

 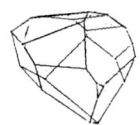

 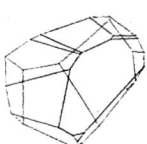

Two polyhedra having random faces. Cross-eye stereograms. Computer graphics by Katsumi Saito. As the total number of faces increases, the number of pentagonal faces also increases.

set of tiles may be chosen to form a fundamental region, it will not be found to be repeated periodically. Voderberg's sea horse and Golomb's sphinx can also be arranged in periodic manners, because each two of their units can be combined in a parallelo-octagon or a rhombus respectively.

Geometers were puzzled whether such "essentially" nonperiodic—in other words, disordered—tessellations could be discovered. However, during a recent period, chaotic because of the Cold War and other conflicts, some essentially nonperiodic tessellations came to be found, though many kinds of units were used. In spite of the Thirty Year's War and other problems, Kepler also left an example, figure ㉙, though the number of kinds of units was unknown.

Penrose, who designed the arabesque of figure ㉟, also tackled the difficult problem. He found two examples, both of which used only two kinds of units, as in figure ㊾. In the upper, the concave quadrangle ADJI and the convex quadrangle BDJI seen in figure ㉗ are arranged so that the J's of both do not coincide, and in the lower, two rhombi, which are similar to ADBI and DHIJ, both of them also seen in figure ㉗, are arranged so that white belts continue and red independent small discs appear.

These tessellations are dominated everywhere by the gold-

Periodic and nonperiodic stackings of golden isozonohedra. A_6's are shown in yellow, O_6's in yellow ocher, B_{12}'s in red, F_{20}'s in dark brown, and K_{30}'s in pink. Each of the left half is derived from only one kind of unit meeting face to face of smaller golden isozonohedra included in each. Each in the right half is made of all the golden osozonhedra, meeting face to face.

Stackings of fist shapes. Each of the upper and lower row in the central section shows both side views of two kinds of units. The upper section of the dark one shows B_{12}'s of one and two frequencies, the left and right sections of both those sections, F_{20}'s and K_{30}'s of one and two frequencies. The bottom row shows examples of nonperiodic honeycombs. The central section shows roofs of single (left) and double (right) layers of Penrose's tessellation seen below in 49.

Japanese traditional paper design "Gourd." Woodcut by Chojiro Senda. Twentieth century. A nonperiodic print of a single wood block is continuously repeated across the whole to produce a striking periodic effect. Such paper, a Kyokarakami, is used as a design of a sliding panel, a Fusuma, in Japanese-style houses.

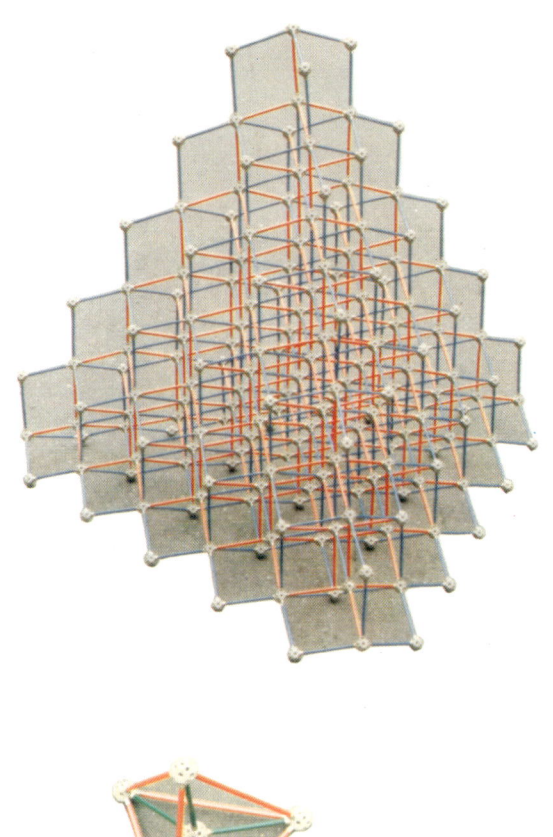

㊼

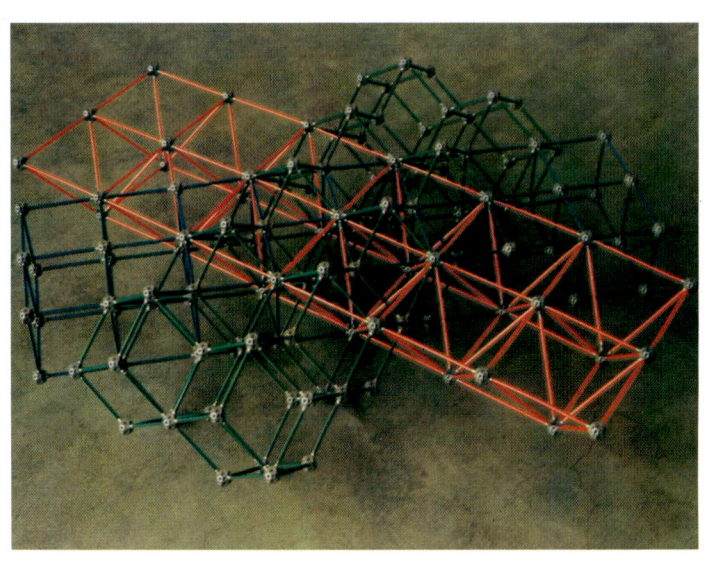

㊽

Golden diamond.

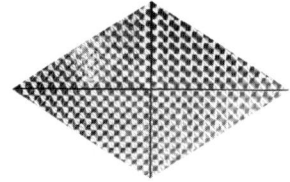

33

Golden isozonohedra. From left to right, A_6 (upper), O_6 (below), B_{12}, F_{20}, and K_{30}.

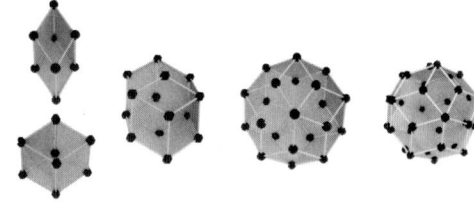

34

Small golden isozonohedra in large golden isozonohedra. From left to right, an A_6 (upper) and an O_6 (below) in a B_{12}, a B_{12} in an F_{20}, an F_{20} in a K_{30}.

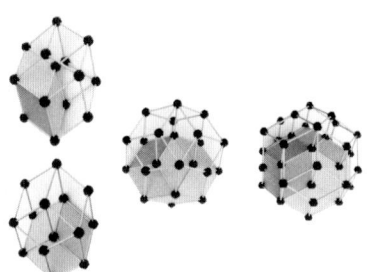

Four-dimensional coordinate axes (upper). Blue lines show positive, red lines negative, directions. Either of them can be embedded in a regular tetrahedron (center) and rhombic dodecahedron (bottom).

Combination of a cubic lattice (blue), octet-truss (red), and rhombic dodecahedral lattice (green). Their vertices coincide with each other.

en ratio. To tell the truth, it seems that there are no essentially nonperiodic tessellations having no relation to the golden ratio.

Recently, some physicists have shown that those tessellations have a practical significance, that they occur in alloys of certain metals, crystallized at certain temperatures. Then, what kind of roofs will be covering this golden town? A stoicheion for them is a rhombus, the Golden Diamond shown in figure 32, whose diagonal lengths have the golden ratio.

There are five and only five convex polyhedra derived from Golden Diamonds, as shown in figure 33: from left to right, two kinds of rhombohedra (A_6 for the upper acute type and O_6 for the lower obtuse type), a dodecahedron (B_{12} after Bilinski), an icosahedron (F_{20} after Fedorov), and a triacontahedron (K_{30} after Kepler).

All are zonohedra and are called the Golden Isozonohedra by Coxeter. These five are intimately related to each other. For example, as shown in figure 34, a B_{12} is obtained by a parallel translation of an A_6 or O_6. Similarly, an F_{20} is obtained from a B_{12}, and a K_{30} from an F_{20}. In other words, if a Golden

Diamond is thought to be equivalent to a two-dimensional square, and an A_6 or O_6 is compared to a three-dimensional cube, then B_{12}, F_{20}, and K_{30} are equivalent to four-, five-, and six-dimensional hypercubes respectively.

From the other view point, a B_{12} is filled by two each of A_6 and O_6, an F_{20} by a B_{12} and three each of A_6 and O_6, and a K_{30} by an F_{20} and five each of A_6 and O_6.

Since they are composed of small golden isozonohedra which themselves fill space, A_6's and O_6's, the golden isozonohedra can fill various portions of 3-space in periodic manners as do those zonohedra, and in nonperiodic manners as do certain nonzonohedra. Figure ⑤⓪ shows examples. Each unit has the constant color. Those in the left half appear when only one kind of unit is used. Those in the uppermost row are face-sharing stackings of A_6's and O_6's. Those of the second row are two kinds of face-to-face period stackings of B_{12}'s, and the third row shows front and rear views of one layer of a face-to-face nonperiodic stacking of B_{12}'s. Those of the fourth and bottom row are periodic (left) and nonperiodic (two on the right) stackings of F_{20}'s and K_{30}'s respectively having faces and O_6's in common.

Each of the right half appears when all golden isozonohedra are mixed, which is possible because they have faces of the same shape. They look like dead flowers in the fall, but each will grow infinitely when suitable golden isozonohedra (especially A_6's or O_6's) are added.

But they are not essential because the periods of parallel translations are also seen. Are there essentially nonperiodic patterns? A fist shape and its mirror image in the central dark section of figure ⑤① are inviting men on the planet into the regions dark in uncertainty. If an A_6 and O_6 are connected with each other across any face, then two kinds of fist-shaped concave decahedra can be obtained, as shown in the upper and lower row in the central section. They are mirror images of each other and one of them looks like the right fist and the other like the left. The right and left figure in each row present their front and rear views.

If two right fists or two left fists are made to shake hands, then a B_{12} is born, as in both sides of the central section of

Modification of Penrose's endless stairway.

Solar house made of golden diamonds. California. Designed by Steve Baer.

the top row. The central one of the same section is a B_{12} of double size.

In other sections, there are an F_{20} and a K_{30} and clusters of double that size. Moreover, a whirlpool spreads and a tornado grows as in the left and right ends of the bottom row.

The central section of the same row shows a roof of one layer (left) and of two layers (right) covering a town whose plan is following one of Penrose's essentially nonperiodic tessellations, shown at the bottom of figure ㊾.

Penrose's golden town has such a roof, though, in contrast to the plan design, it is obviously not essential. Coxeter says that there cannot be any essentially nonperiodic honeycomb. The colors and numerals on the faces of fist shapes were assigned in a search for essentially nonperiodic honeycomb, but the search came to naught.

Scaffolds are necessary to build a roof. What shapes of scaffolds are suitable for the roof of Penrose's golden town? They are not three mutually perpendicular families of lines in 3-space, but six coordinate axes in 6-space, derived from a regular dodecahedron when the body center and each face center are connected. Therefore, figures [33] and [34] have regular dodecahedral joints.

Eight of these regular dodecahedral joints can be put together to form a toroid of Stewart, having one tunnel whose

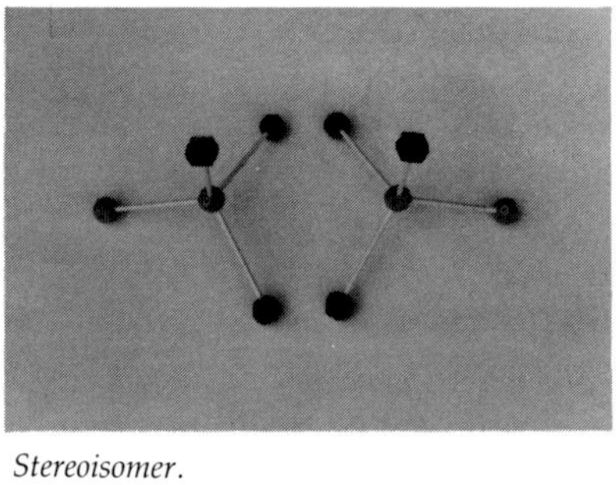

Stereoisomer.

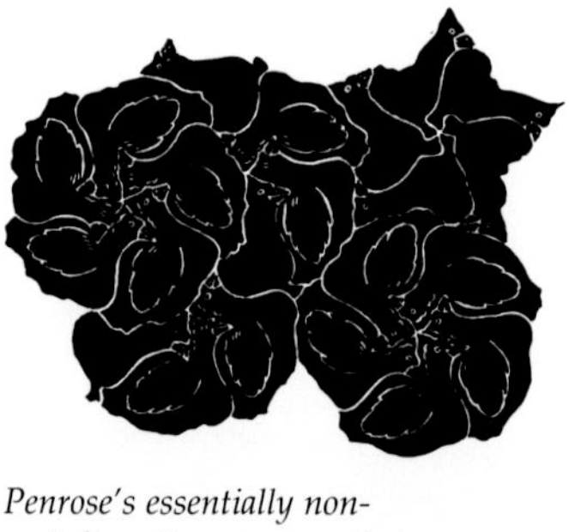

Penrose's essentially nonperiodic pattern composed of hens.

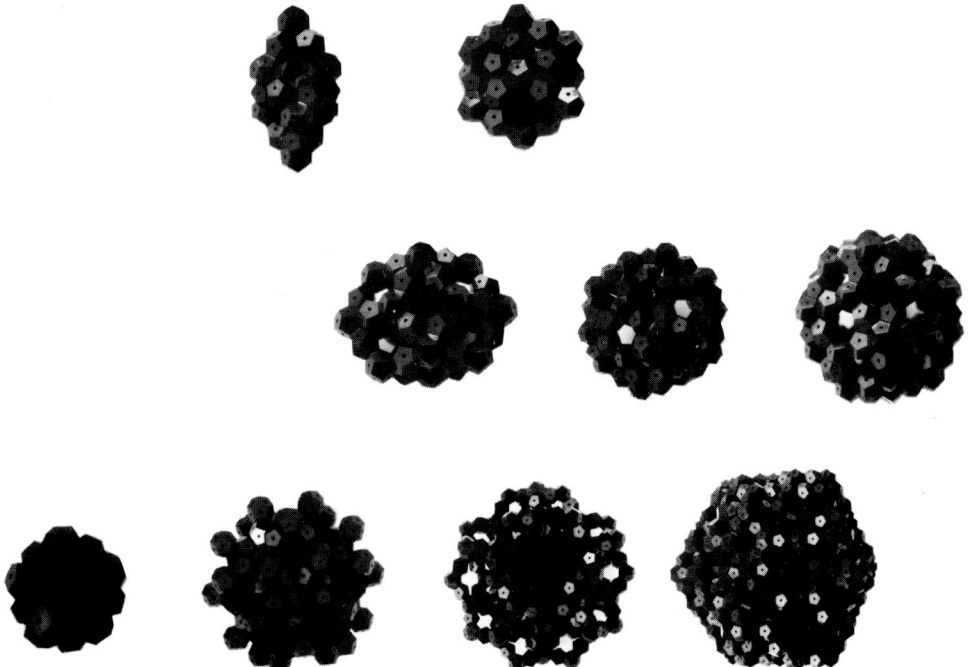

35

Toroids formed from regular dodecahedra. From the left end of the top row to the right end of the second row, an A_6, O_6, B_{12}, F_{20}, and K_{30} appear respectively. The bottom row shows a cell fission of the planet "Golden Diamond" according to the icosahedral symmetry.

outline is a Golden Diamond. The planet "Golden Diamond" can thus be constructed as a toroid of regular dodecahedra. The upper two rows of figure 35 show each golden isozonohedron constructed as a toroid out of regular dodecahedra. In the bottom row of the same figure are various toroids built as cell fissions which grow with icosahedral symmetry. The universe of Plato is nothing but a molecule in the planet "Golden Diamond."

13 Coordinates by Descartes

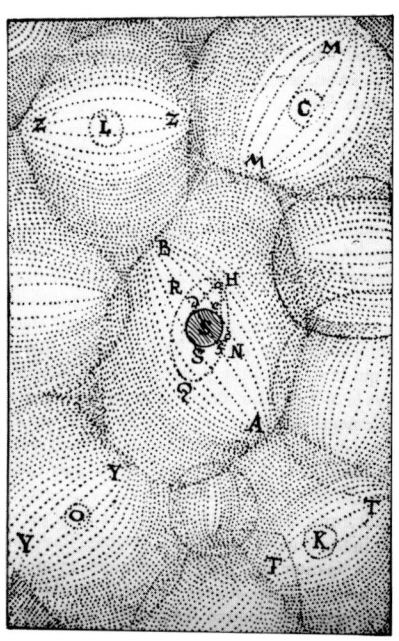

Universe of Descartes according to the vortex theory. From Principia philosophiae.

The universe, which we have shown to be filled with polygons or polyhedra, is also filled with mysterious riddles. Einstein, Fuller, and others intended to clear up the riddles by observing from 4-space and in doing so, greatly contributed to a better future for mankind. When observed from $(n + 1)$-space, n-space reveals its true features and can be treated globally.

The human race, living in 3-space, is well acquainted with events in 2-space and easily recognizes polygons but not polyhedra. The world of polyhedra is born when some two-dimensional planes are combined in 3-space. Polyhedra, therefore, may be regarded as three-dimensional objects from 4- or higher-space.

For this reason, some higher dimensional figures have already been treated in this book. The captivating planet of Golden Diamond has six dimensions. One-sided polyhedra, stellated polyhedra, and so on, need four dimensions to be represented without improper intersections of faces. There were also movable polyhedra, for whom "motion" is an added dimension. To view such a chaotic universe of polyhedra from a higher-space, various coordinate axes should be introduced.

In the fourth century B.C., Aristotle asserted in the opening of his *De Caelo* that there are only three mutually perpendicular directions in our space and that the number "3" is more meaningful than "4" of Plato.

In China, Chuang-Tzu and Chun-Nan-Tzu explained that the universe, "Yü-Chu," contains space, "Yü," having six directions: east, south, west, north, up, and down, and time, "Chu," which extends in one further direction.

In the Christian era, the thinking of Aristotle was handed down by various persons in later ages, such as Ptolemy, J. Buridan, and Galileo. Bacon said in his *Opus Majus* that space has three coordinate axes, as symbolized by the cube or the Trinity, and that time has another axis.

It is usually said, however, that the discoverer of the three orthogonal or oblique coordinate axes ("cartesian" coordinate axes) was Descartes, who conceived of the Vortex theory of the universe. The theory was that whirling dust and dirt fills the universe in a manner quite unlike that in which orderly polyhedra filled the universe of Kepler. It is natural that co-

Face-to-face stacking of truncated icosahedra, that is, soccer balls, according to a rhombic dodecahedral lattice.

ordinate axes seem necessary for such a chaotic Cartesian Universe.

However, Kepler had already used three orthogonal directions in space in order to make clear the riddle of the shapes of snowflakes. He was also aware of a rhombic dodecahedral lattice and even of an octet-truss, mentioned earlier. Decartes, who had read Kepler's *On the Six-Cornered Snowflake*, attached great importance only to the most usual orthogonal axes.

However, a rhombic dodecahedral lattice, which produces a diamond lattice, relates to 4-space, and an octet-truss relates to 6-space, as stated by Fuller. A diamond lattice, which controls the arrangement of carbon atoms in a molecular structure of a diamond, has the shape determined by the four mutually intersected lines seen in a rhombic dodecahedron, as in the bottom of figure ㊾. If the edges of two diamond lattices, having all their lattice points in common, are painted in blue and red, as in the top of figure ㊾, they become a projection of the positive (blue) and the negative (red) coordinate axes in 4-space, into 3-space. The angle between any pair of these axes is 109°28′. If we project along one of the axes, into the plane, we find 120° between the remaining three axes, in projection. In a sense, 109°28′, which is also seen at the center of a regular tetrahedron as in figure ㊾, is the three-dimensional equivalent of a 120° angle in the plane. Modern geometers call this angle the "Maraldi's angle," after two astronomers in the middle of the eighteenth century. In nature, there

Four soap bubbles gathered around a point generating 109°28′S.

Molecular structure of DNA.

are many examples besides a diamond which show this angle: such as real honeycombs, seeds of pomegranate, and the like, as observed by Kepler.

Kumiaki Ito, a modern Japanese polyhedric architect, has found a compound of cubic lattice and octet-truss constructed on the same set of vertices. The former is most suitable for dwelling space, and the latter for the structural members. He has applied the compound to various architectural projects.

Into Ito's compound we may introduce a four-dimensional system of orthogonal axes. It is a rhombic dodecahedral lattice or a diamond lattice on the same set of vertices, as in figure ㊳. In the figure, a cubic lattice is shown in blue, an octet-truss in red, and a rhombic dodecahedral lattice in green.

This diamond lattice appears when the body center and four suitably chosen vertices are connected in a polyhedron having the isosahedral symmetry, and following the lattice, for example, truncated icosahedra (soccer balls) are stacked infinitely face-to-face. This cannot be done if we follow the cubic lattice.

Both the closest and the loosest arrangements of spheres are also related to the diamond lattice and thus to Maraldi's angle. The right angle is rather unusual in art and nature, even in 3-space. Recall how Fuller in *Synergetics* said that a cubic lattice is nothing but an unstable scaffold for a castle in the air invented by mathematicians. He thinks the universe extends infinitely, bounded sometimes by a cuboctahedron, sometimes by a rhombic dodecahedron, which is the dual of the cuboctahedron and has four coordinate axes appropriate to 4-space.

Recently an astronomer said that the planet Uranus may be made of diamond. If so, it belongs in 4-space. The molecular structure of the ice covering Saturn is also ruled by a diamond lattice. Thus there is no need to go all the way to Uranus to see a space of this dimension. We can find 4-spaces even closer to home. Our human bodies are filled with DNA, whose structure is ruled by Maraldi's angles. We have four dimensions even in ourselves.

The Crystal
by Schläfli

14

A 3-space does not stand alone. There are innumerable 3-spaces in 4-space. A component 3-space is called a cell or hyperplane in 4-space. There is an essential difference between a closed polyhedron and a polyhedral cell. The former lacks an inner space, which the latter has. Polyhedra defined by Plato are the former; those by Fuller, a four-dimensional thinker, must be the latter, with their inner spaces.

A four-dimensional figure which corresponds to a polyhedron in 3-space is a polytope or a 4-polytope. It is constructed of polyhedral cells, certain pairs of which have a face in common. A handful of earth, a drop of water, a fragment of fire, a blast of wind, and even a vacuum is such a cell. The three-dimensional universe, which is filled infinitely with them, having faces in common, is a polytope in 4-space.

Hyperdevelopment of the 120-cell.

Among polytopes, the most regular ones are the six regular polytopes which were discovered by Schläfli in the middle of the nineteenth century. Each regular polytope has only one kind of regular polyhedral cell, a fixed number of which fit together around every edge and any two of which have a face in common. They correspond to regular polyhedra in 3-space. The 5-cell, corresponding to the regular tetrahedron, has five regular tetrahedral cells, three of which fit together around some edge. The 8-cell (four-dimensional hypercube), corresponding to the cube, has eight cubic cells, three of which fit together around every edge. The 16-cell, corresponding to the regular octahedron, has 16 regular tetrahedral cells, four of which fit together around every edge. The 24-cell, having no corresponding regular polyhedron, has 24 regular octahedral cells, three of which fit together around every edge. The 120-cell, corresponding to the regular dodecahedron, has 120 regular dodecahedral cells, three of which fit together around every edge. Finally, the 600-cell, corresponding to the regular icosahedron, has 600 regular tetrahedral cells, five of which fit together around every edge.

Of them, the 8-cell is the father and the 16-cell the mother of the 8-cell family, and the 120-cell is the father and the 600-cell the mother of the 120-cell family, by four-dimensional duality. The 5-cell and 24-cell are singletons. Thus there are two "Christs" in 4-space.

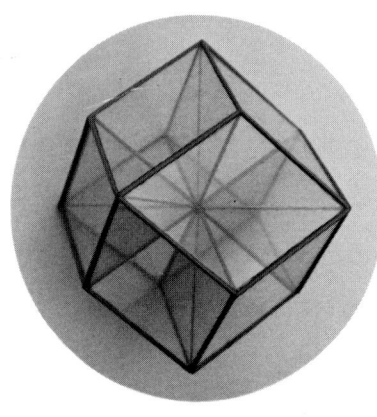

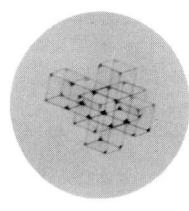

[36]

Hypercube (upper) *and its development as a spatial kite.*

This four-dimensional hypertemari is an analogue of the incense burner shown on page 6. That incense burner is decorated with 12 pentagonal thin blossoms, each of which is inscribed in pentagonal faces of a spherical regular dodecahedron. This being the case, a hypertemari is decorated with 120 regular dodecahedral thick blossoms each of which is fixed in a regular dodecahedral cell of the 120-cell.

It is easy to account for these regular polytopes abstractly, relying on analogy with regular polyhedra in 3-space. However, there are many riddles concerning their concrete shapes, which everyone wishes to solve. The previously mentioned diamond lattice is a stable scaffold on which to build these mysterious figures. When a four-dimensional figure is embedded in this lattice, the role of the right angle in 3-space is played by 109°28′ or by its supplementary angle 70°32′. Thus four cubic cells are changed into rhombohedra, so as to construct a rhombic dodecahedron, as in the bottom of figure ㊾. And if an identical rhombic dodecahedral stacking of four cubes (rhombohedra) is superimposed on the figure ㊾ in such a way that pairs of cubes have rhombic faces in common, the eight cubes form a four-dimensional hypercube.

One three-dimensional projection of a four-dimensional hypercube is as in the upper part of figure [36]. The principal segments, running in four directions, represent the edges. The other lighter line segments, running in three mutually perpendicular directions, simply indicate improper intersections of faces in 3-space. This phenomenon was already seen in the cases of one-sided polyhedra or stellated polyhedra. Two formally distinct vertices coincide at the body center.

The lower figure in figure [36] shows a hyperdevelopment of the 4-cube in which eight 3-cubes are connected face-to-face. Dali painted Christ, nailed to such a cross.

Such a hypercube introduces a new era in model construction. Now we can construct models of the projections onto 3-space of each regular polytope as in figure ㊿. The 5-cell (hypertetrahedron) is shown in the top row. First, by using four green, flattened tetrahedra shown on the left, construct a regular tetrahedron as in the center. Next, envelop them in a blue regular tetrahedron, as on the right. That is all. The result can be represented using edges only as in the central figure of figure ㊾. A four-dimensional water chestnut thus has five vertices. The previously mentioned 8-cell (hypercube) is shown in the second row. First, by using four green, flattened rhomboids shown on the left, construct a rhombic dodecahedron (center). Next overlap four similar blue rhombohedra, upside down, though this is impossible to do in our 3-space. That is all. A four-dimensional die thus has one to eight spherical spots marking its eight rhomboids.

The 16-cell (hyperoctahedron) is shown in the third row. First, on each face of a red regular tetrahedron at the left, put a green regular triangular pyramid whose lateral faces are right triangles, so that a cube having a diagonal on each face appears, as at the second from right. Next, on each face of the cube, put a blue square having two diagonals, thought of as a regular tetrahedron reduced to a plane in which two opposite edges cross at right angles. Then, six newborn edges, which did not appear on the faces of the previous cube, will produce four new green regular triangular pyramids inside the cube. Finally, the innermost cell is a red regular tetrahedron, the same shape as the first, but with reversed order. That is all. A four-dimensional diamond is often mined in this shape.

The 24-cell is shown in the fourth row. First, on each face of a red regular octahedron at the left, put a blue flat octahedron so as to produce a cuboctahedron having two diagonals on its square faces, as at the second from the left. Next, on each square face, put a green square having two diagonals, which is thought of as a regular octahedron reduced to a plane so that two opposite vertices coincide. Then, six newborn vertices, which did not appear in the previous cuboctahedron, will produce eight new blue octahedra in the cuboctahedron. Finally the innermost cell is another red regular octahedron in the reversed order. That is all. There is no corresponding regular polyhedron in 3-space to this 24-cell, but according to Coxeter, this is a hyperrhombic dodecahedron. A four-dimensional pomegranate seed thus has the shape of this 24-cell.

Process of the construction of regular polytopes. From top to bottom, the 5-, 8-, 16-, 24-, 120-, and 600-cell. Each at the left shows the innermost cell and each at the right, the outermost appearance.

Hypercubes appearing in 3 Notations/Rotations Visual Poetry by Toshi Katayama.

Truncations of a 5-cell by hyperplanes which cut the trisect (upper row) and bisect points of edges (lower row).

Geometrical analysis of the face of Kisshoten, a Japanese heavenly maiden. Jyoruriji temple, Kyoto. Twelfth century. At the left is an analysis based on a regular icosahedron, after Matila Ghyka. ABCDEFGH is a regular icosahedron, EF : EG = 1 : 1.618, and IJKL equals a square. If she were a western beauty, O and P would coincide with her pupils, Q the top of nose, B the center of mouth, R the bottom line of jaw, S the top of head, and M and N both sides of her face. The analysis on the right, based on a regular octahedron, is an attempt by the author. ABCD is a regular octahedron, EH : EF = 1 : $\sqrt{2}$, IFGJ a square, and B and D indicate cut lines seen in the sculpture.

90

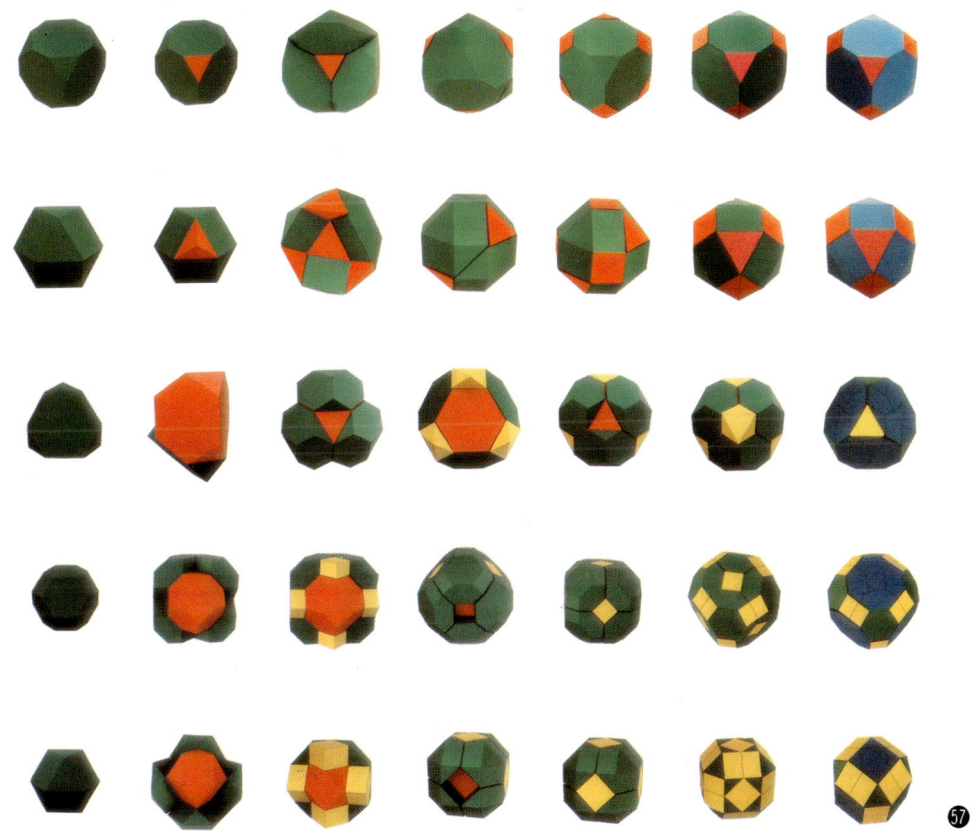

91

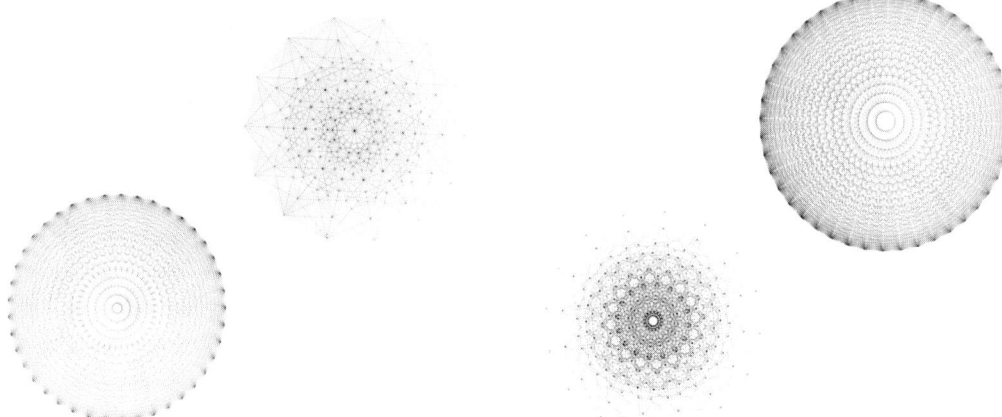

Higher-dimensional regular polytopes. From left to right, 40-dimensional hypertetrahedron, 7- and 8-dimensional hypercubes, and 22-dimensional hyperoctahedron.

◀

Truncations of a hypercube, the 16-cell, and the 24-cell by hyperplanes. The top and second row are truncations of hypercubes through the trisect and bisect points of edges. The third row is a truncation of the 16-cell through the trisect points of edges. The fourth and final row are truncations of the 24-cell through the trisect and bisect points of edges.

◀

Process of the construction of a four-dimensional small stellated dodecahedron. One marked * is the stellation of the central red regular dodecahedron, ● of the second yellow layer, ▼ of the third grey layer, ■ of the fourth grey and green layer, and ◆ of the outermost green and blue layer. The scattered small pieces appear on the way to the hyperstellation.

The 120-cell is shown in the fifth row. First, on each face of a red regular dodecahedra at the left, put a yellow ochre dodecahedron so as to leave 20 dimples. Next, put into each dimple a grey dodecahedron so as to leave 12 new dimples. Further put into each new dimple a green dodecahedron so as to produce a convex 42-hedral outer shape having 12 regular pentagons and 30 parallelohexagons, as at the second from right. Then, on each hexagon put a blue parallelohexagon, which is thought of as a regular dodecahedron reduced to a plane in such a way that two opposite edges coincide. Then newborn edges, which did not appear in the previous 42-hedron, will produce four kinds of dodecahedra from green to red, reversing the order. That is all. This method of construction was already explained in Figure ㊸. A four-dimensional pollen of *Gypsophila elegans* thus has 120 spherical orifices.

Finally, the 600-cell is shown in the bottom row. First, by using 20 red tetrahedra at the left, construct a red regular icosahedral agglomeration, as at second from the left. On each outer face of it, put a yellow tetrahedron, so as to create a star shape as at third from the left. Next, add 30 yellow ochre, 60 grey, 60 dark brown, 60 blue, and 20 white tetrahedra, successively, to produce a convex 72-hedral outer shape having 12 white and 60 blue triangles, as at the second from right. On each blue triangle, put a green triangle, which is thought of as a regular tetrahedron reduced to a plane, so that two faces coincide. Sixty newborn faces, which did not appear in the previous 72-hedron, will produce seven kinds of tetrahedra from white to red, reversing the order. That is all. In it we can find some icosahedral agglomerations of 20 tetrahedra. M. Ghika has shown that the face-proportion of three-dimensional beauty in the western hemisphere relates to a regular icosahedron. If this is so, then the face of a four-dimensional beauty would relate to this 600-cell. She would be rather spherical. By way of contrast, a four-dimensional Japanese beauty would have the symmetry of the 16-cell. This is because it is said that the golden ratio, relating to the regular icosahedron, is found in the beauties in the West, while it seems that in Japan the ratio of $1 : \sqrt{2}$ is found.

However, we would meet only a Japanese-style beauty, and not a Western-style beauty, in 5- or higher-space. In those spaces, there are always three and only three regular polytopes: the n-dimensional regular $(n + 1)$-tope corresponding to the regular tetrahedron, the n-dimensional regular $2n$-tope corresponding to the cube, and the n-dimensional regular 2^n-tope corresponding to the regular octahedron. Unfortunately, we cannot see her unpainted face, for there is no suitable set of coordinate axes, such as the diamond lattice, in 4-space. We are obliged to be satisfied with a rough sketch, or a topological distorted diagram, of her. According to the sketch, an n-dimensional regular $(n + 1)$-tope becomes a regular $(n + 1)$-gon together with all its diagonals, a famous diamond pattern. An n-dimensional regular $2n$-tope becomes a regular $2n$-gon with inner edges which are mutually parallel and have an equal length to the edges of the original $2n$-gon. An n-dimensional regular 2^n-tope becomes a regular $2n$-gon with all its diagonals (except ones which connect opposite vertices with respect to the center).

Figure 55 shows the works of Toshi Katayama, a Japanese-born polyhedric painter living in the United States. A comma-shaped board, which is made of four rhombi colored in red, black, blue, and yellow, is pinned to a regular octagonal board, which is similarly made of eight rhombi. When the comma shape is rotated around the center, a projection of a 4-cube sometimes appears, sometimes disappears. The letters seen here and there on the board form the poem by O. Paz, a poet who was moved by this reversing work. Usually, a poem is written first, and then the illustrations are affixed to it. But in this work, the custom is also reversed.

Shadows such as these, of higher-dimensional regular polytopes, is cast, sharply defined, onto three-dimensional art and nature. Even Fedorov's five parallelohedra, which envelop various three-dimensional designs, are also the mere shadows of higher-dimensional cubes. Each of the parallelohedra in figure 19 is the shadow of a cube, three-, four-, nine-, four-, and five-dimensional, from left to right.

Each shape shown in figure 37 resembles the skeleton of radiolaria, drawn by Haeckel, who walked at an astounding speed, or the skeleton of a geodesic dome designed by Fuller, who talked at an astounding speed. All shapes in 3-space, even Plato and Aristotle, might accidentally be shadows of the only simple figure in higher-space.

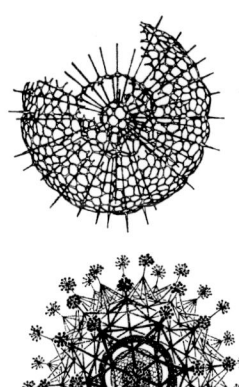

Radiolarias sketched by Ernst Haeckel which show multidimensional appearances.

Process of the construction of the four-dimensional impossible quadrilateral. After Scott Kim. The left end shows a unit and the right, the final appearance.

Four-dimensional postcard (left) and book. Lengths of their three mutually perpendicular edges have the ration $1 : 1.618 : 2.618$ and $1 : \sqrt{2} \div \sqrt[3]{4}$ respectively.

Message from a four-dimensional person. Origamis as hyperhieroglyphs are made by Tokyshige Terada.

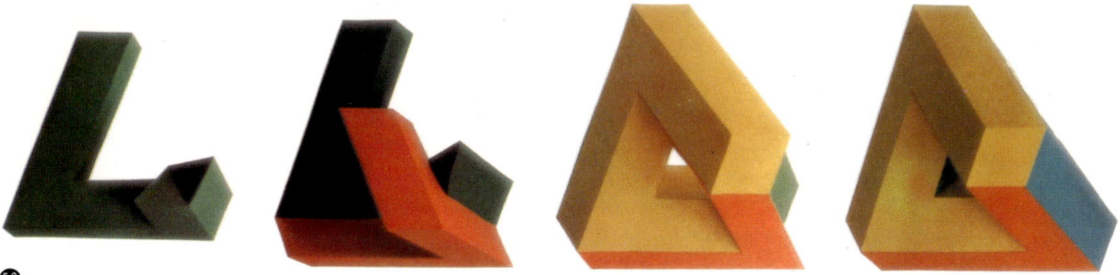

(59)

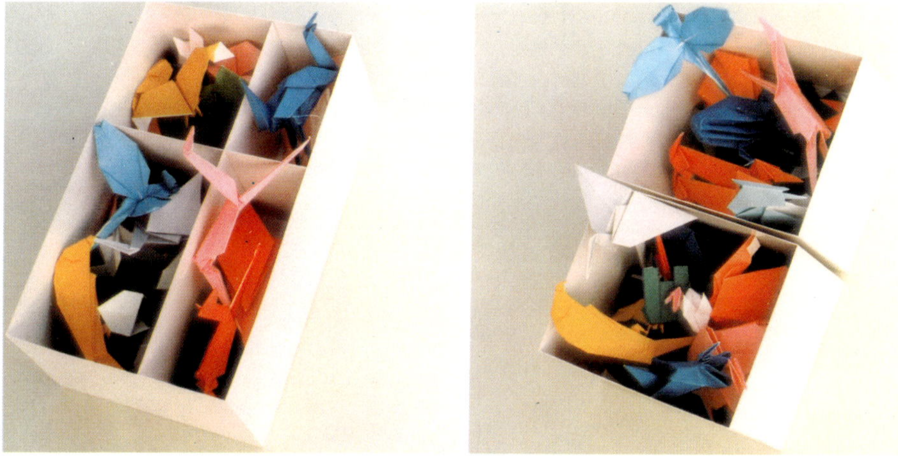

(60)

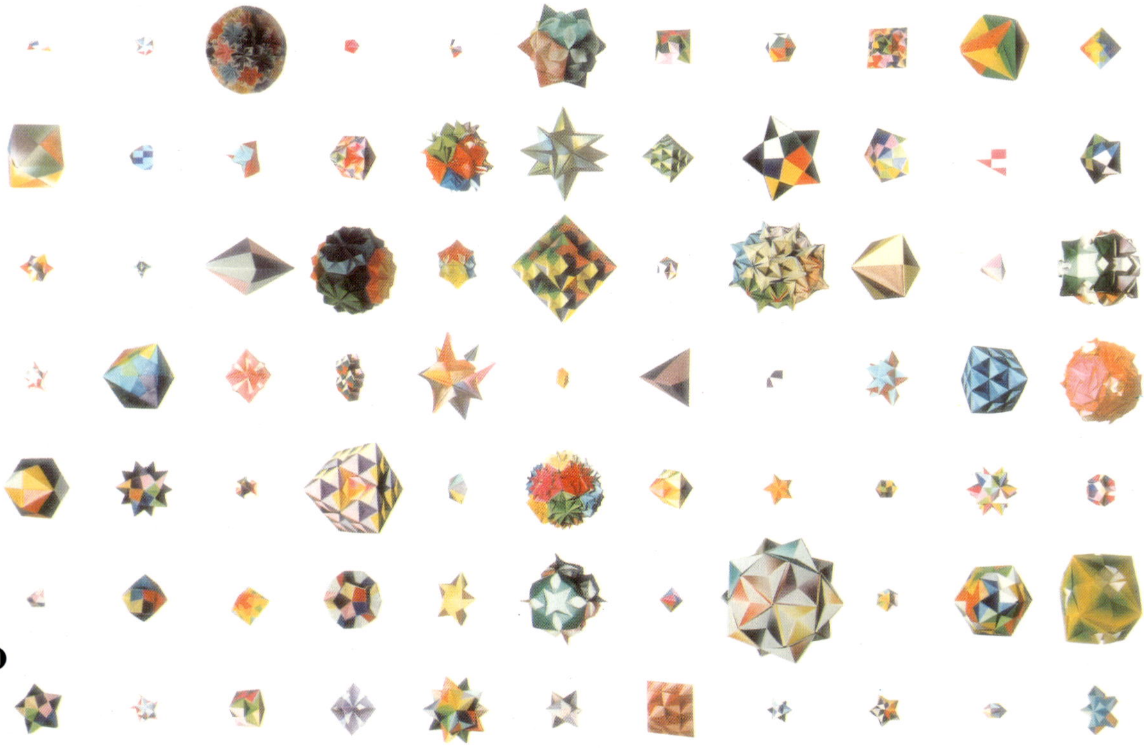

(61)

15 The Creation by Coxeter

Four-dimensional hyper–soccer ball. A soccer ball in 3-space has the shape of a truncated icosahedron whose spherical pentagonal and hexagonal faces are painted in black and white respectively. Similarly, a hyper–soccer ball has the shape of a truncated 600-cell whose icosahedral and truncated tetrahedral cells are inflated in black and white respectively.

The universe changes its form ceaselessly.

The regular 4-polytopes which may make up the future universe must also change their forms. The upper row of figure (56) shows a transformation of the 5-cell in which each vertex is cut off by trisecting the edges by hyperplanes to yield five regular cells and five truncated tetrahedral cells. First, there is a red regular tetrahedron. Around it are stacked four yellow flat truncated tetrahedra. Next, four blue regular triangular pyramids (deformed regular tetrahedra) meet each other at the four corners. Finally, all except parts of the blue pyramids are enveloped by a green truncated tetrahedron.

On the other hand, the bottom row of figure (56) is obtained when each of the edges of the 5-cell is bisected. There are five each of regular tetrahedra (red and blue) and truncated tetrahedra (yellow and green) stacked as in the upper row. This may be a unit of a hyperoctet honeycomb.

Figure (57) shows the truncation of a hypercube and of the 16-cell, which correspond to a cube and to a regular octahedron in 3-space. According to figure [8], figure (57) shows what hyperdiamonds look like. Their four-dimensional coordinate axes form the diamond lattice. The uppermost row shows the truncation of a 4-cube in which each of the edges is trisected. There are 16 regular tetrahedral cells (red) and eight truncated cubic cells (green and blue). The second row shows the truncation of a 4-cube in which each of the edges is bisected. There are 16 regular tetrahedral cells (red) and eight cuboctahedra (green and blue). This may be thought of as a hypercuboctahedron. The outer shape of Fuller's four-dimensional universe may be like this, rather than like a cuboctahedron.

The third row shows the truncation of the 16-cell in which each of the edges is trisected. There are eight regular octahedral cells (yellow) and 16 truncated tetrahedral cells (red, green, blue). This may be thought of as a hyper-Kelvin solid, although it cannot fill 4-space, in a cell-to-cell packing. Curiously enough, if the 16-cell is truncated so that each of the edges is bisected, the resulting figure becomes the 24-cell.

The fourth row of figure (57) shows the truncation of the 24-cell in which each of the edges is trisected. There are 24 cubic cells (yellow) and 24 truncated octahedral cells (red, green, blue). The final row shows the truncation of the 24-cell in

◀ (62)

Explosion of the four-dimensional Platonic universe as the 120-cell.

Four-dimensional hyper–morning glories with leaves. A morning glory in 3-space has the shape of a circular cone whose base is a circular regular pentagon. A leaf has three thin fingers, as becomes a three-dimensional plant. If so, a hyper–morning glory has the shape of a hypercircular cone whose base is a spherical regular dodecahedron. A hyperleaf has four thick fingers, as becomes a four-dimensional plant.

which each of the edges is trisected. There are 24 cubic cells (yellow) and 24 cuboctahedral cells (red, green, blue).

Diamonds glitter deep in the earth and stars shine high in the sky. Figure ⑱ shows the stellation of the 120-cell, not as a supernova but as a hyperstar. This hyperstar corresponds to the small stellated dodecahedron.

Making these models is not so difficult. Each dodecahedral cell, seen in the second row from the bottom of figure ㊴, is replaced by a small stellated dodecahedral cell. Their points intersect each other according to a four-dimensional property. The one marked * shows the stellation of the central red regular dodecahedral cell. The one marked ● is the stellation of the yellow layer, the one marked ▲, the stellation of the grey layer, the one marked ■, the stellation of the grey and green layer, and the one marked ♦, the stellation of the outermost layer. The other small pieces show bits of stardust appearing along the way.

Stars sometimes are hidden by thin clouds, which have the shapes of the two layers of the closest arrangement of spheres in 3-space as shown in figure ⑱. It can be projected onto a plane in such a way that two of the closest circle arrangements overlap each other, so as to put circles of one over the gaps of the other. By analogy, four-dimensional hyperclouds, the closest arrangement of hyperspheres of two layers in 4-space, can be projected into a 3-space in such a way that two of the closest arrangements of spheres overlap each other, placing spheres of one around the gaps of the other. And if the centers of the spheres are connected, the 16-cell is derived. This becomes the coordinate axis system of Fuller's universe.

From such hyperclouds, hypersnowflakes fall. David Brisson, a modern four-dimensional designer, who has already gone into four-dimensional heaven, imagined four-dimensional snowflakes, as in figure ㊳. He imagined three-dimensional snowflakes as shapes shown scattered on the earth of the scene. Each of them is derived from mutually overlapping regular triangles, as is the Star of David. By analogy, hypersnowflakes would be those forms shown scattered in the dark sky in the scene. Each of them is derived from mutually overlapping regular tetrahedra, as is Kepler's stella octangula. The largest transparent flake was made by Brisson himself.

Four-dimensional hyper–red maple leaves. A red maple leaf in 3-space has thin fingers whose number is usually $1 + 2 + 2 = 5$, or sometimes $1 + 2 + 2 + 2 = 7$. If so, a hyper–red maple leaf has thick fingers whose number is usually $1 + 3 + 3 = 7$, or sometimes $1 + 3 + 3 + 3 = 10$, as in the case of a hyperleaf of a hyper–morning glory.

Four-dimensional hyper-thinnest cloud. The thinnest cloud defined by three-dimensional geometers has the shape of two layers of the closest packing of spheres as in 18. Thus hyper-thinnest cloud has the shape of two layers of the closest packing of hyperspheres. One of the layers can be projected into a 3-space as the shape of the closest packing of spheres. Therefore, if two sets of such packings are overlapped so as to put spheres of the one into the gaps of the other, the hyper-thinnest cloud can be derived.

It resembles a cube and appears to be made according to Kepler's description in *On The Six-Cornered Snowflake*.

More precisely, the snowflake should be 4-cube, having as outer boundary a rhombic dodecahedron. In 1983, the space shuttle *Challenger* made a snowflake in space, away from the influence of gravity, following up on a proposal made by two Japanese schoolboys. Although the photo is not clear, the resulting snowflake resembles closely a rhombic dodecahedron. Although Kepler might have already known about four-dimensional snowflakes, as well as three-dimensional ones, his book about them would not be discovered buried under snow. *On The Six-Cornered Snowflake* was in fact buried under dust in a library until it was rediscovered at the beginning of this century. Perhaps someday another book by Kepler will be unearthed, entitled *How Do Pentagonal Hyperflowers Bloom in 4-Space?* Surely they will be like the stellated dodecahedra in figure ㉛ on which buds can be seen.

Four-dimensional hyper-snowflakes by the author. A snowflake in 3-space spreads in a regular hexagon. Thus a hyper-snowflake may spread in a rhombic dodecahedron.

Four-dimensional snowflakes and their three-dimensional analogues. After David Brisson.

16 The Hypersphere by Einstein

Three semicircular canals.

Multidimensional figures of the future have a past, too. Around 1900, many well-informed persons were enthusiastic about multispace or higher-dimensional polytopes, just as those geniuses in ancient Greece or during the Renaissance were enthusiastic about 3-space and polyhedra. One might think that almost all aspects of 4-space, including Einstein's theory of relativity, were exhaustively studied at that time.

Henri Poincaré, a mathematician rumored to be the last many-sided genius, published some best-sellers which discuss multidimensional space. He says in *Science et Méthode:* "It is said that man can recognize three dimensions thanks to three semicircular canals." A mouse in Japan, which has only two semicircular canals, seems to be convinced that he lives in 2-space. If so, it is possible that four-dimensional beings, having four semicircular canals, really exist. They may, however, be puzzled by an "impossible quadrilateral" in 4-space which was devised by S. E. Kim, a modern artist, by generalizing the famous "impossible triangle" in 3-space.

The impossible triangle is a two-dimensional figure which is thought to be impossible to realize in 3-space. Similarly, the impossible quadrilateral is a three-dimensional figure which is thought to be impossible to realize in 4-space. It can be constructed in 3-space using four L-shaped congruent blocks as in the bottom row of figure (59). The final outer shape shown on the right end looks like a "fat" skew quadrilateral whose "edges" are four elongated regular hexagonal prisms, a kind of outer shape of four-dimensional cuboids. However, at least one of these cuboids cannot be realized without distortion, even in 4-space.

Impossible triangle.

The impossible triangle is suggestive of the famous Möbius strip; its discoverer, A. F. Möbius, was the very person who first explained the concept of four dimensional figures. Möbus said, in a short paper, that the right and left hands may coincide in 4-space. Some say that the study of four-dimensional figures started with that observation, in 1827.

Shortly before that, however, Kant had already described multidimensional spaces in various treatises. In his first thesis, he already emphasized the necessity of a geometry of multispace. His famous *Prolegomena* included an inquiry into the difference between right and left. The philosophy of Kant, though difficult for us three-dimensional persons, would be simple for four-dimensional beings.

About 100 years earlier, in the middle of the seventeenth century, J. Wallis said if there were four-dimensional beings, they would have an appearance more curious than the Chimera or the Centaur. His contemporary, H. More, explained that spirit might be four dimensional, permitting spirit and body to penetrate each other without any difficulty. In spite of More, Leibniz doubted that two bodies in 3-space could freely interpenetrate like two shadows on 2-space. More was a contemporary of Kepler and Descartes, and his thought came to resemble rather closely that of the latter, with its emphasis on the dualism of spirit and body. Lonely Kepler, who was ignored by Descartes and Galileo, and estranged even from More, would probably have been absorbed in 4-space, had the concept only assumed a clearer form in his age. He had already found a rhombic dodecahedron, an outer shape of a four-dimensional hypercube. Further, he was able to openly defend the heliocentric system of Copernicus, by observing

Chimera.

from the three-dimensional universe from the outside as if he were God.

On the God of the Middle Ages, 4-space had cast its shadow. Dante says in the "Paradiso" of *Divina Commedia*: "God appears everywhere, everytime, beyond any restriction." He appears in many mirror images in the universe. The laws of nature cannot hold true in heaven because in heaven one can see everything in the same scale, and with the same clarity, no matter how far away it is. This may explain the four-dimensional scenery.

Such a scene was already imagined by Plato. It has been rumored that Plato was the very first person who intuitively knew 4-space. By the famous allegory of a cave explained in *The Republic*, he declared that this actual 3-space is only a shadow of a 4-space.

Plato says: There is a dark cave having a flat wall on which two-dimensional shadows of three-dimensional objects are projected by a candlelight. If some newly born infants are brought up in this cave so that they cannot see things other than the two-dimensional shadows, then they will grow up thinking that they live on a flat land. One day, when they come of age, one of them will be tempted to go out of the cave. His friends in the cave will feel anxious that if he goes out of the cave and sees the mysterious "three-dimensional" space, he will lose his reason. Despite their fears, he is forced to leave the cave and to see the three-dimensional world. Instantly, he is astonished by not a two- but a three-dimensional spectacle. He rushes into the cave to tell his partners about the splendid three-dimensional world. But his partners do not trust him and tell him haughtily that he has been made crazy by seeing a curious thing, the "three-dimensional" world.

Analogously, this three-dimensional world may also be a shadow of another curious "four-dimensional" world.

Plato seemed to like "four," not "three." Why? It might be because of the Pythagorean tetractys, in which a solid (cell) is represented by four circles. The universe would have had four dimensions even if we had not been taught so by Einstein.

The ancient Egyptians, who lived another 2000 or 3000 years before Pythagoras, may also have been aware of the fourth

Fuller's cornucopia.

dimension. As Ghyka tells us, the ancient Egyptians applied various cuboids, which relate the golden number 1.618 to the outer shapes of something meaningful. Among them, the golden solid, the ratio of whose edge lengths is $1 : 1.618 : 1.618^2 (= 2.618)$, has a beautiful property and was used in a king's chamber. If this solid is divided into four cuboids so that one of them becomes a cube, then one of the remaining, having an edge in common with the cube, becomes a golden solid. The golden rectangle, whose curious and beautiful features were previously shown, has a corresponding property. However, it takes a three-dimensional person to appreciate its beauty. Just so, the beauty of the golden solid can be appreciated by four-dimensional, not three-dimensional, persons. The model of the golden solid is not so beautiful to 3-persons as the name would suggest, but might be so beautiful for 4-persons that they would wish to use it as the standard form for a four-dimensional canvas or postcard, as shown at the left of figure ⑥⓪. (Recall that a three-dimensional canvas or postcard usually has the shape of a golden rectangle.) The part in which a yellow bat is placed forms a cube and the part in which a pink deer is placed forms a golden solid.

The right of the same figure shows two cuboids as two pages of a four-dimensional book, the ratio of whose edge lengths is $1 : \sqrt[3]{2} : \sqrt[3]{4}$. No matter how repeatedly one of them is divided into halves, the edges of newborn cuboids have a similar ratio. Such a property corresponds to that of the $\sqrt{2}$-rectangle, the ratio of whose edge lengths is $1 : \sqrt{2}$. No matter how repeatedly this rectangle is divided into halves, the edges of newborn rectangles have the same ratio as the original. Therefore, the sizes of almost all books, newspapers, and other paper in 3-space are standardized as the A or B proportions of this rectangle.

In any case, hyperhieroglyphics are modeled by origamis, folded paper, as in Japanese culture. They could serve as four-dimensional letters, because they look like one-sided polyhedra representing four-dimensional features. Figure ⑥① shows polyhedral origamis forming a four-dimensional sentence. A traditional "Kusudama" which is seen in the third from the left at the top row of figure ⑥①, looks like Fuller's cornucopia.

Qui.

A 3-person cannot read the hieroglyphics because we cannot understand a four-dimensional landscape. We are, so to speak, four-dimensional blind. A message from 4-space might have reached Japan without our knowing it.

Also in China, the word "Yü-Chou" was used in the third or second century B.C. to mean space and time and Zenlike dialogues about the multidimensional universe were prevalent. Chuang-Tzu, a philosopher at that time, says: "Chi," a fictitious animal having only one leg, is envious of a centipede who can move on a line with so many legs. The centipede envies a snake who can sweep smoothly on a plane without legs. The snake is jealous of the wind, which can fly freely in 3-space. The wind is envious of an eye, which can see many things without moving. And the eye envies the mind, which can understand everything without looking. Chuang-Tzu may have intended to say that "Chi" is 0-, the centipede 1-, the snake 2-, the wind 3-, the eye 4-, and the mind 5-dimensional.

Onto such thinking of the ancient Chinese or Japanese, Plato might cast his shadow, as previously mentioned. According to his *Timaeus*, the three-dimensional world at that time was surrounded by the regular dodecahedron. If so, the present and future universe floating in four-dimensional time-space must have the shape of the 120-cell.

Kepler says in his *Misterium Cosmographicum* that the zodiac should be divided not into 12 but rather into 120, the latter having the least common multiple of the numbers 3, 4, 5, 6, 8, 12, 20, and 30, each of which is a number of vertices, edges, or faces of a Platonic solid.

Also the most basic pattern of Fuller's geodesic domes is the spherical 120-hedron, as at the right of figure ③. The 120-cell, which embraces 120 regular dodecahedral cells, is "the least common mutiple solid" of the universes of Plato, Kepler, and Fuller.

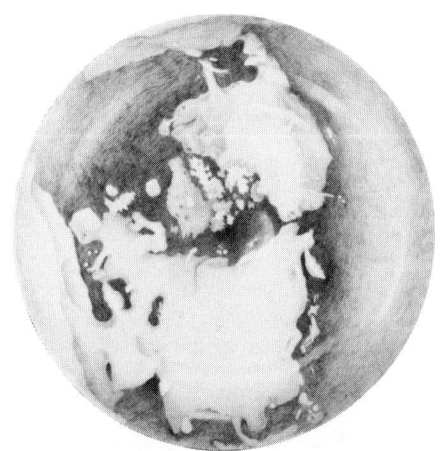

Four-dimensional hyper-terrestrial globe. The terrestrial globe in 3-space is projected onto 2-space so that thin leaves, continents or islands, scatter in a circle, becoming linear shapes near the edge. If so, the hyper-terrestrial globe can be projected into 3-space so that thick fruits, four-dimensional continents or islands, scatter in a sphere, becoming planar near the surface.

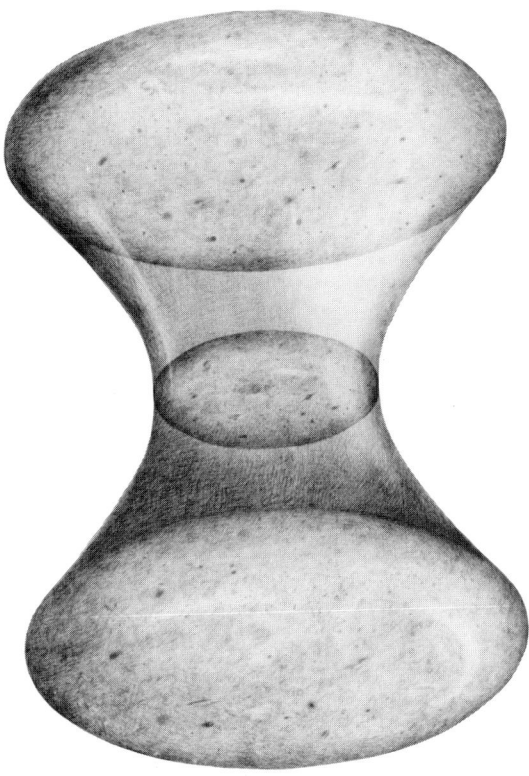

Shape of the universe defined by Einstein. It is represented by the equation $X^2 + Y^2 + Z^2 - T^2 = K^2$, and is projected into 3-space so that an ellipsoid moves, changing its volume from bottom (the past) to center (the present) and to top (the future) along the U-axis. This shape is an analogue of the hyperboloid of revolution of one sheet seen on page 56.

Furthermore, in the 120-cell one can inscribe five other regular polytopes in such manner that all vertices coincide with vertices of the 120-cell.

If the future universe will explode, it would be split into 120 of Plato's cosmos as in figure ⑫, and fly away into a black hole, shining as in figure ㊾.

Everything ends in a four-dimensional black hole.

Bibliography

Abbott, E. A., *Flatland,* Dover, 1952.
Aristotle, *De Caelo.*
Baer, S., *Zome Primer,* Zomeworks Co., 1970.
Ball, W. W. R. and Coxeter, H. S. M., *Mathematical Recreations and Essays,* University of Toronto Press, 1974.
Barbaro, D., *La Pratica della Perspettiva,* 1569.
Barratt, K., *Logic and Design,* George Godwin, 1980.
Baston, V. J. D., *Some Properties of Polyhedra in Euclidean Space,* Pergamon, 1965.
Beard, C. R. S., *Patterns in Space,* Creative Publication, 1973.
Beck, A., Bleicher, M. N. and Crowe, D. W., *Excursions into Mathematics,* Worth Publishers, 1969.
Brisson, D. W., *Hypergraphics,* Westview, 1978.
Burt, M., *Spatial Arrangement and Polyhedra with Curved Surfaces and Their Architectural Applications,* Technion, 1966.
Chandler, D., Grünbaum, B. and Sherk, F. A., *The Geometric Vein,* Springer, 1981.
Coxeter, H. S. M., *Introduction to Geometry,* John Wiley & Sons, 1965.
———, *Twelve Geometric Essays,* Southern Illinois University Press, 1968.
———, *Regular Polytopes,* Dover, 1973.
———, *Regular Complex Polytopes,* Cambridge University Press, 1974.
Coxeter, H. S. M., Du Val, P., Flather, H. T. and Petrie, J. F., *The Fifty-Nine Icosahedra,* Springer, 1982.
Critchlow, K., *Order in Space,* Thames and Hudson, 1969.
———, *Islamic Patterns,* Thames and Hudson, 1976.
Cundy, H. M. and Rollett, A. P., *Mathematical Models,* Oxford University Press, 1961.
Dürer, A., *Unterweisung der Messung,* Verlag Walter Uhl, 1972.
Ehrenfeucht, A., *The Cube Made Interesting,* Pergamon Press, 1964.
Fuller, R. B. and Marks, R., *The Dymaxion World of Buckminster Fuller,* Southern Illinois University Press, 1960.
Fuller, R. B., *Ideas and Integrities,* Macmillan, 1963.
———, *Operating Manual for Spaceship Earth,* Southern Illinois University Press, 1969.
———, *Synergetics,* Macmillan, 1975.
———, *Synergetics 2,* Macmillan, 1979.
Gardner, M., *The Ambidextrous Universe,* Basic Books, 1964.
Gheorghiu, A. and Dragomir, V., *Geometry of Structural Forms,* Applied Science, 1978.
Ghyka, M. C., *The Geometry of Art and Life,* Dover, 1977.
Grillo, P. J., *Form, Function and Design,* Dover, 1960.
Haeckel, E., *Art Forms in Nature,* Dover, 1974.
Hale, N. C., *Abstraction in Art and Nature,* Watson-Guptill, 1980.
Heath, T. L., *A History of Greek Mathematics,* Clarendon Press, 1921.
Heninger, S. K., *The Cosmographical Grass,* The Huntington Library, 1977.

Hilbert, D. and Vossen, C., *Anschauliche Geometrie*, Verlag von Julius Stringer, 1932.
Hinton, C. H., *The Fourth Dimension*, Swan Sonnenschein, 1904.
Holden, A., *Shapes, Space, and Symmetry*, Columbia University Press, 1971.
Joly, L., *Structure*, Editions Idea Suisse, 1975.
Kepes, G., *Module, Proportion, Symmetry, Rhythm*, George Braziller, 1965.
———, *Structure in Art and in Science*, George Braziller, 1965.
Kepler, J., *Mysterium Cosmographicum*.
———, *Harmonices Mundi*.
———, *The Six-Cornered Snowflake*, Oxford University Press, 1966.
Koestler, A., *The Watershed*, Doubleday, 1960.
Lalvani, H., *Transpolyhedra*, Lalvani, 1977.
Loeb, A. L., *Space Structures*, Addison-Wesley, 1976.
Lysternick, L. A., *Convex Figures and Polyhedra*, Dover, 1963.
Macgillavry, C. H., *Fantasy and Symmetry*, Abrams, 1965.
Manning, H. P., *Geometry of Four Dimensions*, Dover, 1914.
———, *The Fourth Dimension Simply Explained*, Dover, 1960.
Munari, B., *Il Quadrato*, 1960; *Il Cerchio*, 1964.
Pacioli, L., *De Divina Proportione*, Bunryu Reprint, 1973.
Pauling, L. and Hayward, R., *The Architecture of Molecules*, W. H. Freeman and Co., 1964.
Pearce, P., *Structure in Nature Is a Strategy for Design*, MIT Press, 1978.
Pearce, P. and Pearce, S., *Polyhedra Primer*, Van Nostrand Reinhold, 1978.
Plato, *Republic*.
———, *Timaeus*.
Popko, E., *Geodesics*, University of Detroit Press, 1968.
Pugh, A., *Polyhedra*, University of California, 1976.
———, *An Introduction to Tensegrity*, University of California, 1976.
Schattschneider, D. and Walker, W., *M. C. Escher Kaleidocycles*, Ballantine Books, 1977.
Senechal, M. and Fleck, G., *Patterns of Symmetry*, University of Massachusetts, 1977.
Slothouer, J. and Graatsma, W., *Cubics*, The Cubic Constructions Center, 1970.
Snyder, R., *Buckminster Fuller*, St. Martin's Press, 1980.
Steinhaus, H., *Mathematical Snapshots*, Oxford University Press, 1969.
Stevens, P. S., *Patterns in Nature*, Little, Brown & Co., 1974.
———, *Handbook of Regular Patterns*, MIT Press, 1980.
Stewart, B. M., *Adventures among the Toroids*, Stewart, 1970.
Thompson, D. A. W., *Growth and Form*, Cambridge University Press, 1968.
Toth, L. F., *Regular Figures*, Pergamon, 1964.
Wachman, A., Burt, M., and Kleinmann, M., *Infinite Polyhedra*, Technion, 1974.
Wells, A. F., *Three Dimensional Nets and Polyhedra*, John Wiley & Sons, 1977.
Wenninger, M. J., *Polyhedron Models for the Classroom*, N. C. T. M. Publication, 1966.
———, *Polyhedron Models*, Cambridge University Press, 1971.
———, *Spherical Models*, Cambridge University Press, 1979.
Weyl, H., *Symmetry*, Princeton University Press, 1952.
Williams, R., *Natural Structure*, Eudaemon Press, 1972.
Wolf, K. L. and Wolff, R., *Symmetrie*, Böhlau-Verlag, 1956.
Wong, W. *Principles of Two-Dimensional Design*, Van Nostrand Reinhold, 1972.
———, *Principles of Three-Dimensional Design*, Van Nostrand Reinhold, 1977.
Wood, D. G., *Space Enclosure Systems*, The Ohio State University, 1973.
Wulff, J. *Structure*, John Wiley & Sons, 1964.
Yale, P. B., *Geometry and Symmetry*, Holden-Day, 1968.
Zalgaller, V. A., *Convex Polyhedra with Regular Faces*, Consultants Bureau, 1969.

Afterword

It is a most delightful event for me that this unusual book by Koji Miyazaki, the foremost polyhedric graphician in Japan, has just been published. The plan for the writing of this book began in conversations we had with the Asakura Publishing Company, while working on the manuscript for our book *Figures and Projections*, published by Asakura in 1979.

Since I was lecturing on graphics at Kyoto University, I was able to involve myself in the project from the beginning. Soon afterwards, I moved to Kyushu University, at a great distance from Kobe, and I regret that I was no longer able to follow the book's progress. Now the work sees the light of day; I thank the author and publishers from my heart for their endeavors.

Koji Miyazaki is a scientist, geometer, artist, and even something of a philosopher on polyhedra. Some idea of his scientific interests are already visible in the above mentioned book, *Figures and Projections*. However, *An Adventure in Multidimensional Space* is unique, showing the very essence of his thinking and creation. It seems that his true love is for the universe filled with polyhedra.

Colorful models of polyhedra made by Miyazaki and by his collaborators have always fascinated me. His dogmatic observations on the history of polyhedra will please his readers. He attaches rather more importance to creative imagination than to scientific reliability. I am convinced that this book will contribute to areas of art, architecture, graphics, geometry, and science.

MICHIO MAEKAWA
Professor Emeritus of Kyoto University
Professor of Kyushu University

Afterword

It is my lasting memory that when I was 13 years old, my elder brother made a parallelopiped for me to submit as part of my homework in mathematics. This polyhedron, all of whose faces are parallelograms, was so mysterious to me—and also to my teacher—that my mark in mathematics was AAA! My teacher told me that my ability in geometry was wonderful and that I should become an architect. A little later, I became a student of architecture.

My distinct memory is that, when 21 years old, I designed a distorted pentagonal dodecahedral concert hall, as an exercise in architectural design at the university. This shape, like a polyhedral clamshell, was so mysterious to my teacher that my mark in architectural design was CCC! My teacher said to me that my direction in architectural design was crazy and that I should become a geometer. A little later, I became a teacher of descriptive geometry at a university.

The third great incident in my life, at the age of 42, is the publishing of this book, in which parallelopipeds and distorted pentagonal dodecahedra play the leading roles.

As is well known, polyhedra have literally many planar faces which can be made of paper. Therefore, if you begin to take an interest in them, you will wish to make models using colored paper. Your models will have a striking appearance, and you will wish to preserve them forever. Soon your room will have no space to store them because of their large volume. They are like empty boxes, or "Kara bakos" in Japanese. Usually we take them apart after taking color photos. Such colorful photos have piled up around me in disorder and chaos during the years I have taught descriptive geometry.

In this chaos, Professor Michio Maekawa, a historian of architecture who can appreciate the values of both past and future, found a cosmic universe. It is from his suggestions that the outline, contents, and title of this book were born.

Surely the birth of this book resembles the birth of the universe, out of chaos. It needed, of course, the support of many other persons, including Professor Masahiko Yokoyama, a historian of science at Kobe University, to construct the space ark "Polyhedra," which now starts out on its voyage to the depth of the universe. The names of these persons are mentioned in this book, insofar as possible. Above all, in imitation of Plato, Kepler, and Fuller, I express by gratitude to all of the universe filled with polygons, polyhedra, and polytopes.

KOJI MIYAZAKI

Index

A_6, 81, 83
A-module, 41, 52
Andreni's honeycomb, 41, 51
Annular torus, 54
Archimedean anti-prism, 22, 24
Archimedean dual, 22–23
Archimedean prism, 22, 24, 41
Archimedean solid, 16, 22, 24, 25, 41, 69

B_{12}, 81–84
B-module, 41, 52
Body-center, 6, 52, 86
Bucky's Bubble, 34, 45

Cartesian coordinate axis, 84–85
Catalan's solid, 22
Cell, 87–89, 96, 102
Chiral regular compound, 64
Circle arrangement, 97
Circumcircle, 72
Circumsphere, 6
Closed polyhedron, 54, 87
Closest packing, 41, 45–49, 51, 98
Cloud, 49
Concave polyhedron, 54
Convex polyhedron, 4, 23, 24, 45, 54–55, 60, 80
Convex polyhedron having regular faces, 25, 45
Cosmic cup, 25, 28, 29–32, 33, 40, 64
Covering, 2
Cube, 4, 17, 20, 38, 41, 44, 48, 51, 52, 53, 57, 64, 67–68, 88–89, 98, 103
Cube family, 16, 21, 24, 45, 69–70, 71, 72
Cubic lattice, 17, 41, 45, 57, 80, 85, 86
Cuboctahedron, 16, 21, 39, 45, 49, 51, 53, 66, 86, 89, 96
Cyclic symmetry, 9

Decasect, 4, 8
Deltahedron, 23, 24, 25
Diagonal plane, 66, 89
Diamond lattice, 85, 86, 88, 93, 96
Dihedral angle, 4, 54–55
Dihedral symmetry, 9
Dodecahedron family, 16, 21, 24, 45, 69–70, 71, 76
Dual polyhedron, 21, 22, 48
Dymaxion sky-ocean map, 33

Edge, 13, 28, 33, 34, 45, 51, 52, 54, 64, 66, 68, 71, 73, 88, 89, 92, 93, 100, 103, 104
8-cell, 87, 88, 89

Elongated rhombic dodecahedron, 51
Essentially nonperiodic tessellation, 76, 77–80, 84
Euclidean space, 56
Euler's equation, 54

F_{20}, 81–82, 83, 84
Face, 13, 22–24, 33, 49, 52, 54, 55, 57, 61, 66, 69, 71, 72, 77, 81, 87, 88, 89, 92, 96, 104
Finite polyhedron, 54
5-cell, 87, 88, 89, 96
4-polytope, 87, 96
Fuller dome, 34
Fuller's cornucopia, 102, 103

Geodesic dome, 33, 34, 45, 93, 104
Geodesics, 29
Geodesic sphere, 28, 29, 31, 34
Golden Diamond, 80–81, 82–83, 84–85
Golden isozonohedron, 77, 80, 81, 83–93
Golden ratio, 69, 70, 73, 80, 92
Golden rectangle, 73, 103
Golden solid, 103
Golomb's sphinx, 76–77
Great circle, 4
Great dodecahedron, 49, 61–64
Great icosahedron, 61, 64
Great rhombicosidodecahedron, 16
Great rhombicuboctahedron, 16, 21
Great stellated dodecahedron, 49, 61–64

Hexagram, 69
Hexasect, 4, 8
Higher-dimensional regular polytope, 90, 100
Honeycomb, 2, 41, 51, 66, 70, 86
Hyperbolic paraboloid, 56
Hyperboloid of revolution of one sheet, 56, 105
Hypercircular cone, 97
Hypercloud, 97
Hypercube, 87, 88, 90, 96, 101
Hypercuboctahedron, 96
Hyperdiamond, 96
Hyperflower, 98
Hyper-Kelvin's solid, 96
Hyperleaf, 97
Hyper–morning glory, 97
Hyperoctahedron, 88
Hyper–octet honeycomb, 96
Hyperplane, 87, 89, 90, 96
Hyper–red maple leaf, 97

Hypersnowflake, 97, 98
Hyper–soccer ball, 96
Hypersphere, 97, 98, 100
Hyperstar, 97
Hypertemari, 88
Hyperterrestrial globe, 104
Hypertetrahedron, 88
Hyperthinnest cloud, 98

Icosahedral symmetry, 9, 19, 83, 86
Icosidodecahedron, 64, 67
Impossible quadrilateral, 93, 100
Impossible triangle, 100–101
Infinite polyhedron, 54, 55, 57
Insphere, 6
Intersphere, 6

Jitterbug, 49

Kelvin's solid, 51, 52
Kepler-Poinsot solid, 64
Kiriko, 39
Klein bottle, 66
Kusudama, 103

Labyrinth, 57
Line, 2, 3, 6, 66, 70, 72, 76
Lobachevski-Bolyai space, 56, 57
Loosest arrangement of sphere, 44, 86

Maraldi angle, 85–86
Möbius strip, 64–66, 100

n-dimensional regular $(n + 1)$-tope, 93
n-dimensional regular 2^n-tope, 93
n-dimensional regular $2n$-tope, 93
Nonperiodic honecomb, 77, 84
Nonperiodic stacking, 73, 77, 81
Nonperiodic tessellation, 76–80, 82

O_6, 80–81, 83
Octahedral symmetry, 9
Octa-sect, 4, 8
Octet honeycomb, 17, 44, 51, 52, 53
Octet-truss, 17, 45, 49, 57, 64, 80, 85, 86
109°28', 85–86, 88
120-cell, 87–88, 89, 92, 96, 97, 104–105
120°, 85–86
One-sided heptahedron, 66
One-sided polyhedron, 57, 66, 84, 88
Open polyhedron, 54
Origami, 57, 64, 93, 103
Orthogonal regular compound, 64

Packing, 2, 4, 32
Parallelogon, 52
Parallelohedron, 51–52
Pentagonal dodecahedron, 73
Pentagonal number, 3
Pentagram, 13, 61, 69
Periodic honeycomb, 77, 84
Periodic stacking, 73, 77, 81

Periodic tessellation, 77
Platonic solid, 8, 22, 38, 39, 51
Polycube, 17, 44, 51, 53, 64
Polygon, 2, 8
Polyhedron, 4, 7, 23, 29, 33, 34, 39, 51, 54–55, 60, 64, 68, 70, 71, 86, 100
Polylink, 66–67
Polytope, 87
Prism, 4, 22
Pseudo-rhombicuboctahedron, 16, 22, 24, 25, 41
Pyramid, 4, 17
Pythagorean tessellation, 2, 3–4, 7, 18, 45, 56

Quanta module, 44, 52–53, 69

Regular compound, 64
Regular dodecahedron, 6–7, 8, 12–13, 20, 21, 22, 24, 38, 39, 44, 45, 48, 51, 60, 61, 64, 69, 70, 71, 72, 73, 83–84, 87, 89, 92, 97, 104
Regular 4-polytope, 87, 96
Regular hexagon, 45, 56, 69, 98, 100
Regular hexahedron, 4
Regular icosahedron, 6–7, 8, 9, 16–18, 21, 29, 38, 44, 60, 61, 64, 73, 87, 89, 92–93
Regular n-gonal prism, 6
Regular n-gonal pyramid, 6
Regular octahedron, 6, 8, 9, 17, 18, 20, 21, 38, 44, 49, 51, 57, 60, 64, 66, 87, 89, 93, 96
Regular pentagon, 13, 24, 41, 44, 61, 69, 70, 71, 72–73, 92, 97
Regular polygon, 2, 4, 16, 21, 22, 24, 32, 36, 55, 64, 69
Regular polyhedral cell, 22
Regular polyhedral sphere, 6
Regular polyhedron, 4, 6, 8–12, 16–24, 35, 39, 41, 52, 54–56, 60, 64, 87, 89
Regular polytope, 25, 88, 89, 93
Regular sponge, 55–57
Regular tessellation, 2, 7, 8, 56
Regular tetrahedron, 4, 6, 8, 9, 16–18, 20, 22, 38, 39, 44, 45, 49, 51, 52, 64, 80, 85, 87, 88, 89, 92, 93, 96, 97
Rhombic dodecahedral lattice, 45, 57, 85, 86
Rhombic dodecahedron, 44, 48, 53, 56, 67, 80, 85, 86, 88, 98, 101
Rhombicosidodecahedron, 16
Rhombic triacontahedron, 44, 48, 64, 66
Rhombicuboctahedron, 16, 21, 22, 24
Rhombohedron, 52, 80, 88
Rhombus, 70, 77, 80
Riemannian space, 56
$\sqrt{2}$-rectangle, 103
Rotational symmetry, 6

Saddle polyhedron, 45, 57
Saddle surface, 57
Semiregular polyhedron, 22, 23, 41, 54, 64
Semiregular tessellation, 2
Simple polyhedron, 54
600-cell, 87, 89, 92, 96
16-cell, 88, 89, 90, 92, 96, 97
Small rhombicosidodecahedron, 16
Small rhombicuboctahedron, 16, 21
Small stellated dodecahedron, 49, 61–64, 90
Snub cuboctahedron, 16, 21
Snub icosidodecahedron, 16
Solid, 3, 23

DISCARDED